# The International Library of Ethics, Law and Technology

## Volume 17

More information about this series at http://www.springer.com/series/7761

Ana Delgado
Editor

# Technoscience and Citizenship: Ethics and Governance in the Digital Society

Springer

*Editor*
Ana Delgado
Centre for the Study of the Sciences
and the Humanities (SVT)
University of Bergen
Bergen, Norway

ISSN 1875-0044 ISSN 1875-0036 (electronic)
The International Library of Ethics, Law and Technology
ISBN 978-3-319-32412-8 ISBN 978-3-319-32414-2 (eBook)
DOI 10.1007/978-3-319-32414-2

Library of Congress Control Number: 2016957507



Printed on acid-free paper

This Springer imprint is published by Springer Nature
The registered company is Springer International Publishing AG
The registered company address is: Gewerbestrasse 11, 6330 Cham, Switzerland

# Foreword

*In this era of rapid and sweeping advancement, we see the old world struggling to guide and restrain the process of advancement into the new (next?) world. Recording companies howl bloody murder in the old courts about people "stealing their livelihood" by making and distributing pirate copies of their intellectual properties. Yesterday's telephone companies become today's facilitators of information and entertainment access. World governments gnash their teeth at the possibility of new technologies sparking sweeping economic change and the dashing of the old world's entrenched economic power structures. Change will happen according to the will and abilities of the masses, regardless of the old world sensibilities.* Rockpiler, TECHNOLIFE forum participant.

The politics of emerging science and technology are truly complex, meaning that there is a multiplicity of legitimate perspectives on the issues relating to what they are, what they should be, and by whom and for whom they should be made or, more accurately perhaps, allowed to come into being as ways of life. In the politics of science and technology, many things are happening at the same time, comprising a multitude of interests, power plays and debates, temporalities, decision processes, actors and institutions in government, private enterprise and civil society. Part of what is going on can be grasped by tracing the multiple actors' interests, say, to understand the intervention of certain industrial corporations into political debates. However, conventional interest-based explanations – explaining acts and decisions in terms of the supposed interests of the actors – are of limited value when we enter the politics and governance of contemporary, or future, science and technology. This is so for at least two reasons. First, the perception of the ambiguous character of some of the developments – the thrill and the terror, as it were – is shared by many actors, also scientists, engineers and policymakers. Hence, allegedly rational projects coexist happily with people's most basic hopes and aspirations, wants and needs. Second, and related, what we are dealing with here are largely things that do not exist other than as plans, hopes, visions and imaginations. The contemporary techno-political innovation scapes and agendas, in all their breathless fluidity and ferment, are still poorly understood as political and social phenomena. However, one general point is clear: these discourses of promise and even of anxious *demand* from the coming future have increasingly colonized the future as what should have been an imagined open landscape of collective possibility and collective will.

The opening quote above is taken from the TECHNOLIFE project's deliberative forum on *converging technologies and the future of the human body*. The participant is grappling with and articulating many of the issues that also concern policymakers, scientists and scholars of science and society: Which are the new sociotechnical structures emerging in this 'era of sweeping and rapid advancement'? What's going on in current techno-driven economies? What *should* be going on? Is there any place left for at least some vestiges of democratic politics? Or should the institutions of old simply be swept away, as corrupt purveyors of illusion and empty posturing? The project was conducted against a backdrop of policy processes in Europe and the European institutions that increasingly appeared to be out of touch with 'European publics and citizens'. This was problematic for a number of reasons, significantly since the very same citizens were also routinely invoked as the main sources for legitimating the introduction of these technosciences, including as their correspondingly needy users and consumers. The problem indeed looked similar to the contemporary landscape of economic policies, strongly driven by supply-side economics of credit and consumption, with only poor or non-existent consideration given to the demand side (let alone need or distributional sides) of the equation. Hence, there was a perceived gap, or lack, in public understanding, according to which the publics, citizens and prospective users must be brought into innovation lest Europe 'lose out in the global competition'. However, the question remains as to what extent we really understand these (non-)relations of the elites and their imagined publics with actual publics and users.

The argument could be made that these imagined publics of European elites and their private public relations consultants never really existed outside of this more-than-normally politically insulated EU policy elite's imaginations, including those select few entrepreneurs and visionaries enlisted to feed into the policymakers' innovation agendas. Indeed, research into public understandings has demonstrated again and again how the typical 'public' has never trusted 'science' in any unqualified and universal manner; publics have always exercised a pragmatic and realistic sense of science's fallibility and its proneness to exaggerate what it knows and can control. Thus, there appears never to have been the golden age from which came the fall into generic public mistrust. Rather, the growth of the apparent overall range and energy of public dissent from dominant policy and scientific commitments on particular issues (and not generically but discriminatingly) may have much more to do with the huge intensification of the number, pervasiveness and intensity of the normative demands which are being made by such technoscientific and commercial innovation concerns, for more and more public subordination to those demands, which have rarely if ever been allowed to become matters of public debate and vote, in a timely manner. This blind-spot corresponds with the tendency for conventional thinking and practice in 'public engagement with science' to problematize the publics who may be involved, but not the usually ill-defined 'science' and technoscientific agendas with which those publics are meant to engage.

In this sense and others, Rockpiler and other participants confirmed to us that the problem was not so much with 'science' as with the institutions and power relations within which science was given its dominant meanings. An important further part of this 'public' problem both for the policy institutions and actors (including so-called 'independent' scientists and commercial interests and 'authorities') is just this resistance to recognizing that what is invoked as 'science' by this EU (and leading member-states) policy culture , as if it were a singular coherent agent of public authority, is in reality a shallow and incoherent, even self-contradictory ragbag of opportunistic reactive references, in an ultimately self-harmful attempt to shore up European (and member-state) government viability, vis-à-vis their publics. Approaching a quarter century ago now, I referred to this (Wynne 1993) as a need for institutional reflexivity when it comes to the problems arising in increasing policy uses of science for purposes of attempted public persuasion and authority. Those needs still exist – but with only marginal improvements, so too unfortunately do the institutional self-delusions. It is a matter for discussion as to the extent to which these deep blind-spots are sustained more by scientific institutional anxieties and habits, or by political (and science media) ones, and how these mutually reinforce one another.

Thus, while one typical public participant, Rockpiler, as above, explicitly identified those institutional blind-spots, it remains doubtful whether his views could ever be taken seriously by those institutions that were allegedly asking for his opinions. The diagnosis that the problem is just as much with the institutions and meanings as with the 'science' itself (whatever precisely was meant by this reference) was indeed an old insight, which can be traced back to the early days of science and society scholarship. However, as implied by Rockpiler as well as the diverse contributions to this book, that insight may come with some twists of its own in this age of technoscientifically pervasive digital media. In the remainder of this book, we elaborate on this intuition of the TECHNOLIFE project, as it appears across different fields of technoscientific innovation, and as it appears to different scholars, mainly working within the fields of science and technology studies and ethics. It remains a troubling and frustrating experience that while we encounter individuals amongst the prevailing institutional policy culture who are well able to articulate these observations, despite their diffuse recognition in this world, and despite their prime importance, they never appear as *public record* issues for policy and its commercial patrons and beneficiaries to address, in an accountable way.

The initial idea for this book originated in a meeting of the TECHNOLIFE project that took place in Barcelona in 2010. TECHNOLIFE was a European Commission–funded project whose goal was to develop novel methods of engagement of selected publics to produce 'alternative ethical' frameworks for emergent technologies. By emergent technologies we had in mind a new kind of technology from the 'charismatic megafauna' technologies of classical controversy studies, which could be defined and identified in well-focused ways – and which also arguably allowed the exaggerated expert belief that their risks and impacts could also be predicted. By emergent technology we meant an increasingly prominent if varied *kind* of technology: mainly and perhaps always now *informed* by digital technologies; more difficult to locate; thus, less clearly bounded; more decentred; more versatile and flexible; and also more amenable to diverse user innovation and adaptation.

Although like most STS scholars we would insist that all technologies, even relatively fixed and centralized ones such as that ultimate of charismatic megafauna, nuclear power, are both quintessentially social, and emergent or indeterminate in significant degree, these more pervasive and prominent forms of *emergent technology* of more recent provenance can be distinguished by these foregoing general characteristics.

The TECHNOLIFE project included studies of, and deliberations about, a large range of such emerging technologies, as described and analysed in the following chapters. When first discussing this book, the idea of the members of the project consortium was that it should provide an overview of these technologies in a way that would be accessible for a broad audience, including Master students in subjects such as STS and ethics, but also (we hoped) interested/concerned citizens, NGOs, policy actors and so on. This eventually influenced the style in which the book is written, combining a moderate academic registry or idiom with rich empirical descriptions. This general style of the book corresponds with the spirit of the TECHNOLIFE project, which aimed at engaging in broad and critical dialogue on a set of such emerging technologies. Our goal then was not to uncover and identify representative voices among the citizenry, but rather to provide a room for those voices that are commonly less represented but who might have valuable insights about the broader predicaments of our 'knowledge societies'. As we affirmed through this avowedly experimental collective attempt, this more ambitious and more complex goal also required a quite fundamental shift in prevailing understanding of the technical, and the social, and of their interrelations and mutualities. This also implies a different politics, as yet to be defined.

Lancaster University
Lancaster, UK

Brian Wynne

# Contents

# Introduction: Technoscientific Governance – The Ethics and Citizenship of New and Emerging Technologies

The reader of this book will be familiar with, and perhaps even dazed by, recent 'technoscientific events'. Some of these events have an air of science fiction, things we could have only imagined 10 years ago. In 2013, Facebook exceeded 1 billion active users. Facebook is a global platform for exchanging information and connecting people, a 'social network'. It is also a reservoir of data, a collection of individual biographies and a type of collective digital memory. When users die, their Facebook identities remain as ghostly memorials. Like other social networks, Facebook has installed itself into the everyday lives of citizens. However, the intimate aspect of this technology mixes disturbingly with its global ambitions: When Facebook tried to go public in the stock market, what was truly being sold? Facebook, Twitter and similar online platforms have created doubt regarding what is public and what is private. These platforms have also served as channels with which to share concerns and enable movements to grow, such as the Arab Spring and the 15 M Movement in Spain. Somewhere between science and fiction, Wikileaks' Edward Snowden wanders in airports around the world and iris scanners track people, indications that something may be changing. In tangible manners, such emerging technologies enable new forms and relations between body and mind, the manner in which we experience one another, our governments and even the spaces we inhabit. Digital maps are used by governments to control the masses in the event of natural catastrophes; however, such maps can also be used by citizens to design the spaces in which they want to live. This book suggests that there is something special, perhaps unique, about these types of emerging technologies as to the ways in which they interfere with politics. The book addresses technoscience as it enters the domains of the human body, the movement of citizens and space by providing descriptions at the intersection of technology, politics and ethics.

In their analysis of current technological changes, some authors have suggested a type of shift from modern science to current technoscientific research and innovation (Carrier and Nordmann 2010; Bensaude-Vincent et al. 2011). Although acknowledging that scientific enterprise has always been technological (Haraway 1991; Rheinberger 1997; Latour 1999) and oriented towards industrial production, Carrier and Nordmann emphasized that in the age of technoscience, research is increasingly driven by the ambition to radically transform nature and society. As Latour and other STS (Science and Technology Studies) scholars noticed some time ago (Latour 1993; Shapin and Schaffer 1985), modernity as an epoch counted on a number of devices and mechanisms to create the illusion that science and politics could work separately. Science would produce *pure facts* that are not polluted by values and interests. Whereas modern science and technology have been legitimized by claims of truth and progress, emerging technosciences such as nanotechnology and geo-engineering are predicated on the promise of novel and wonderful designs that will radically change people's lives: nanorobots that, inserted into the body, will help to fight disease and an atmosphere that will behave according to designed plans. Perhaps more explicitly than ever before, nature and society have become the objects to be transformed by technological interventions in the pursuit of a desired future. Innovation works on the limits of what can be imagined. With the focus on novelty and transformation, scientific insistence on pure fact has become somewhat relaxed (Delgado 2016; Rommetveit and Wynne forthcoming).

Instead of portraying our times as a break with modernity and modern styles of governing science, this book identifies an increasing tendency towards an innovation-driven manner of conducting research that can be traced back to the period following World War II. Since that time, scientific research has become more explicitly targeted towards economic growth (see Chap. 1). The governing of science has followed those developments, specifically with increasing public investments in large-scale research programmes that are technology-driven. Largely, the spreading of digital infrastructures, platforms and tools enables a new 'engineering' mode of research that has 'design' as a core practice. In many ways, the digital era enables new types of technoscientific objects that exist somewhere between the technical and political intervention. 'Privacy by design' in the emerging field of biometrics and the visualization of future natural catastrophes related to climate change are but two examples of such emerging technoscientific objects. In Europe, the faith in technological solutions to control and transform nature and society informs the Horizon 2020 programme. Technological innovation is portrayed as the key to coping with European 'grand challenges' such as an ageing population, energy independence and climate change (JIIP 2012). Because scientific policies are (perhaps more than ever) invested with the hope that desired futures can be created, scientific research is oriented towards specific 'contexts of application' (Carrier and Nordmann 2010).

This book focuses on three sets of emerging technosciences: human enhancement technologies, biometrics and GIS (Geographical Information Systems). By including each of these themes as a thematic section, the book provides some insights into how the human body, space and the atmosphere reappear as territories to be reexplored and recomposed in the pursuit of promising technological interventions and transformations. With a primary focus on governance, the book provides insights into institutional developments as well as the public responses that accompany particular processes of technological emergence. Governance, as the encompassing theme of the book, favours a particular perspective of how technologies might enter people's worlds to produce change. The book addresses processes of technological emergence and provides empirical descriptions of how transformations of citizens' lives are *already occurring* as part of such occurrences. This perspective will enable a resituating of technological emergence and its governance in the present in an attempt to balance a policy tendency to orient the analysis of scientific governance towards the future. Finally, the book provides a particular approach to governance by combining STS and ethics. The life-world of citizens as intervened by technology is considered from this double approach to remind the reader that issues of institutional power, inclusion and exclusion simultaneously reveal ethical dilemmas and tensions.

## Technoscientific Governance: Ethics and the Life World of Citizens in a Digital Society

The notion of governance that this volume proposes is introduced in Strand and Funtowicz's chapter (Chap. 1). These authors revisit some key historical moments in the movement from the foundations of modern science to current technoscientific governance. According to Strand and Funtowicz, in a European context, government evolved into governance in science policy as an intuitional attempt to regain legitimacy in the face of an increasing public distrust of governments and expert institutions (see also Wynne 2006; EC 2010). In such a context of crisis in representative politics, governance was proposed as a programme of 'broadening and improving government' by more public participation (see p. 6). Whereas accountability, transparency and inclusion have been presented as elements of good governance, ideals have not always matched realities (Delgado and Strand 2009). After the mid-1990s and the beginning of the 2000s, Europe experienced a participatory 'boom' (Delgado et al. 2011); however, the 'new politics of talk' (Irwin 2006) was created with the goal of producing consensus although dissenting voices were often misrepresented. Somewhat paradoxically, a crisis of political representation was meant to be addressed by constructing a representative public (Lezaun and Soneryd 2007).

During all those years of political discussion, institutional representations of the public have coexisted with non-invited or spontaneous public reactions and developments (Wynne 2007). Environmental movements flourished in connection with nuclear and GM (Genetically Modified) technologies, raising important concerns that were misrepresented by governments. Arguably, through activists' campaigns and other forms of collective action, such movements have endeavoured to be visible and have their views and values included in the democratic politics of representation. However, the spreading of digital infrastructures and related emerging technologies across society may be creating a different experience of what it means to live with technology (see Chap. 2) and perhaps new manners of articulating political action and citizenship (Rommetveit and Wynne, forthcoming). Arguably, formations such as DIY (Do It Yourself) medicine, the open hardware movement or the citizens' groups that use GIS technologies act as users and as producers, appropriating and modifying the technologies themselves. These groups are a type of public who are perhaps less insistent on realizing their claims to institutions (as 'old' social movements) and who articulate their concerns by 'doing'. In this light, citizens as users of technologies, and particularly digital technologies, can be perceived as public articulations of another type. They do not necessarily expect that their concerns and claims should be addressed by institutions (in a Deweyan sense).[1] Rather, these citizens perform more direct forms of action, becoming involved with the technologies in novel manners. Immediate realities such as the human body or the local space are the concerning issues to be directly addressed, beyond institutional arrangements and constraints. These sort of interventions pose new challenges to classical expert-lay divisions. In neoliberal politics, large state infrastructures tend to dissolve while representative politics weakens. Technology as applied to health (for example, in personalized medicine) and other realms of everyday life increasingly becomes a private matter, a matter of individual choice. A key question is how and to what extent the new formations of the public are challenging such technological individualization, proposing new forms of collective action and citizenship.

On the side of governments, new forms of constructing publics are emerging along with the spreading of digital tools and infrastructures. Digital platforms and design software are crucial technical developments enabling the proliferation of a new institutional talk regarding 'social innovation'. In the Horizon 2020 framework, the public should be actively involved in technological development, as co-designer or end user; public values should be integrated in the early stages of scientific research; and responsibility for the production of desired technological trajectories should ideally be shared (Owen et al. 2012). In a new politico-technical constellation that is innovation-driven, the governance of science has been configured in terms of distributed responsibility. A relevant question is to what extent the 'social contract' and 'technological citizenship' (Frankenfeld 1992) might enable an empowerment of the citizenry or rather a liberation of the states from previously demandable liabilities.

Emerging technologies such as geo-engineering, biometrics and converging technologies for human enhancement are oriented towards changing realities in the pursuit of desirable futures. At a governance level, this orientation towards the future results in a paradoxical situation that has been designated the Collingridge dilemma: 'The social consequences of a technology cannot be predicted early in the life of the technology. By the time undesirable consequences are discovered, however, the technology is often so much a part of the entire economic and social fabric that its control is extremely difficult. This is the dilemma of control. When change is easy, the need for it cannot be foreseen; when the need for change is apparent, change has become expensive, difficult and time-consuming' (Collingridge 1980:11 in Nordmann 2010). At the base of this dilemma is the radical uncertainty that is inherent in scientific research (Strand 2002). Whereas scientific practice has attempted to suppress uncertainty by a number of black-boxing epistemic strategies

[1] (Dewey 1927).

(Latour 1987), emerging technosciences attempt to control natural and social uncertainty by design strategies (Nordmann 2007). At a governance level, a manner in which to address the paradoxical situation noticed by Collingridge is developing strategies to build social capacity so that *better* decisions are made in the present and *better* technological trajectories are developed (Guston 2014). In the face of uncertainty, ethics have been introduced to the governance of technoscience. Both in Europe and the US, research on the ethical aspects of emerging and ICT-based technologies[2] is becoming mandatory in funding research programmes. Another crucial reason why ethics is emphasized as an important element of technoscientific governance is because emerging technologies are perceived as potentially having a direct effect on the life of citizens, transforming them in substantial manners. Such technologies may have a double nature: being promising but also intimately intrusive.

In the FP7 and in the framework of the Science-in-Society programme, the European Commission launched a call to develop 'new ethical frameworks' for new and emerging technologies. One of the research projects resulting from that call was TECHNOLIFE.[3] From 2009 to 2012, the authors of this book collaborated to develop that project. Using audiovisual material, TECHNOLIFE provided an online forum for discussion of three technological domains: digital maps, technologies of body enhancement and biometrics (corresponding to three sections of this book). Citizens who were concerned or whose lives could be affected by those technologies were invited to discuss the social, ethical and technological implications of those technologies. The project was designed to allow dissent rather than produce a 'representative' public opinion. A final movie provides a summation of the forum's discussions and concludes, 'It is not easy to sum up our experience. Many responses were highly imaginative and elaborate, and only a few examples could be presented here. Many emphasized that we are living in times of great change. No clear ideologies or political platforms could be singled out. Many saw politics and bureaucracies as outdated institutions. Many were critical of the monopolies of the mass media and large corporations. Some saw free software, open sources and an open Internet as indicating more sustainable futures'.[4] As Strand and Funtowicz emphasize in their chapter, a diagnosed problem was that 'The potential is particularly large when the speed of science-based innovation is not matched by the speed of institutions' and people's capacity to cope with the change. The result is a particularly poorly governed (as government and governance) process of innovation possessing characteristic difficulties as a fast-moving, unpredictable, uncontrollable and sometimes invisible target'. Such criticism was communicated to the Commission, comprising (or so we thought) a reflective and positive experience of inclusive governance.

---

[2] As the 'Science-in-Society' programme becomes integrated into scientific programmes in the FP7 of the European Commission.

[3] TECHNOLIFE stands for 'Transdisciplinary Approach to the Emerging Challenges of Novel Technologies: Lifeworld and Imaginaries in Foresight and Ethics'.

[4] www.technolife.no

## Governing Citizens' Bodies, Movements and Space(s): The Chapters of the Book

In different manners, the chapters of this book address technological change, focusing on how technologies insert themselves into the lives of citizens, transforming people and creating new political and ethical configurations and problems. The contents of the book are organized into four thematic sections. The first section introduces the notions of governance, ethics, publics and technological change that structure the book. The first chapter has an introductory character with a general focus on the governance of emerging technologies. In this chapter, Roger Strand and Silvio Funtowicz focus on the ethical and political complexity inherent in emerging technologies by introducing some key issues and by situating such technologies in a historical perspective. Complexity largely arises from the emerging and radically uncertain nature of these new technologies. The technologies are ethically ambiguous; however, they may produce great transformations. In addition, these technologies largely exist only in 'plans, hopes and visions'. A multiplicity of legitimate perspectives may be included in decisions regarding technological developments; however, decisions are inevitably made as technologies are emerging, at which time interests themselves are developing as the technologies emerge. A question that orients this paper is why researchers (such as the authors themselves) should be interested in proposing new ethical frameworks for the governance of emerging technologies. Researchers in the fields of science and technology studies and ethics, the authors argue, are active partners in the processes of sociotechnical change and should reflectively consider their roles.

Chapter 2 is authored by Kim Jepsen, Ana Delgado and Margareta Bertilsson and addresses technological change, focusing on how publics may come into being in contexts of technological emergence. Public concerns regarding technologies emerge as people attempt to accommodate 'stranger' technologies in their everyday lives. The chapter analyses how those concerns emerge as communal and they are articulated by collective action. The chapter emphasizes that 'issues unfold and are articulated in multiple and contested ways, on the grounds of shared concerns. Some concerns and issue articulations might remain at the margins of representation, evoking a public spectre' (p. 28). In this manner, the chapter focuses on the political nature of the public and anticipates a notion of citizenship that is grounded in the idea of the 'common' (and the community). This motif reappears throughout the book.

The second section of the book investigates how emerging technologies might transform the materiality and meaning of the human body and mind. The focus of the section is technologies of 'enhancement', that is, technologies of body modification that are targeted to enhance performance and life conditions. From cosmetic surgery to bionic implants and software for enhancing the activity of the brain, technologies of body enhancement are quite heterogeneous. However, body enhancement technologies share at least three important features. First, these technologies assume and challenge a certain notion of 'normality'. Second, as objects of enhancement, the human body and mind ambiguously appear as both malleable and limiting. Ironically, in their limitations, the mind and body offer unlimited possibilities for technological transformation. Third, as the technologies themselves, such transformations of bodies and minds remain emergent, and in most cases, these technologies remain no more than visions and hopes. Although technologies of body modification have existed since ancient times, the ambition to transcend the limitations of the human mind and body (as in transhumanists' visions) remains largely within the realm of fiction (Delgado et al. 2012). Thus, this section includes a final chapter on science fiction narratives of body enhancement. In this chapter, Miquel Barceló and Louis Lemkow draw on a rich collection of stories, presenting an account of the diversity of human enhancement technologies and the diversity of ethical and political dilemmas that may arise with them. Science fiction narratives are presented as a medium with which to explore technologies that remain in development (or are simply imagined) and whose implications cannot be grasped in any manner other than the speculative imagination. By presenting imagined technological landscapes, science fiction narratives address classic ethical and political questions such as: 'What does it mean to be human? Who should have control over technology? How are people affected by science and technology?' (p. 68). The chapter attends to a number of controversial enhancement technologies such as genetic enhancements and introduces a special focus on the implications of digital technologies. An identified classic theme in science fiction narratives is the uncoupling of the body and mind. Such uncoupling would, for example, enable the creation of a multiplicity of copies of the same person or allow people's minds to be uploaded to computer memories. Issues of identity and immortality occupy a central position within science fiction narratives in addition to questions of how the technological enhancement of individuals may trigger issues of representative democracy and 'human rights' and ultimately produce social inequality and injustice.

Similar themes reappear in Chap. 4 in relation to the academic literature. Authored by Søren Holm, this chapter outlines the primary ethical controversies related to body modification and enhancement and identifies the factors that create such controversies. An initial controversy addresses the limits of the human body and its normative status. Against such limits, transhumanist claims to the right of 'morphological freedom' are emerging. These claims raise ethical and political questions regarding whether freedom should be the master value of democratic societies. A related controversy regards the possible implications of unequally distributed body modifications and enhancements. In terms of social justice, 'freedom of morphology' may result in increasing the divide between the rich and the poor. Finally, this chapter discusses ethical problems related to extreme transhumanist visions that foresee immortality and the overcoming of all human limits by technological convergence. Chapter 4 focuses on what it means to be 'normal' and on the dynamics of inclusion and exclusion that may emerge with the development of body enhancement technologies. Chapter 3 returns the focus to politics and governance, relating this chapter to the previous section. In this chapter, Kim Jepsen introduces the notion of *citizenship*, situating the concept in a context of technological emergence. This chapter provides sociological analysis of how deaf people have experienced radical changes in their lives after acquiring a bionic implant. In a sense, such implant recipients have become cyborgs. With a body that has been technologically enhanced, these people remain strangers to socially accepted notions of what is to be a human being. Drawing on ethnographic materials, this chapter explores 'unresolved issues of what it means to live as citizens with technological innovation' (p. 36). The chapter examines how the human body as the domain of technological interventions becomes an object of science, government and industry. In their everyday experience and attempts to live with their bionic implants, such citizens redefine their social identities. Inserted into the everyday lives of citizens, technology influences both the emergence of new communities (with a renewed sense of belonging) and new forms of social exclusion. The notion of citizenship that this chapter proposes addresses the political dynamics of inclusion and exclusion, suggesting manners in which to address technologically mediated diversity.

The third section of the book addresses the emergence and implementation of technologies for tracking and controlling citizens' movements. This section focuses on how biometric technologies are being used by governments in the European Union (EU), creating a new area of 'security' while, many argue, placing basic elements of what composes citizenship in liberal democracies at risk. The chapters in this section emphasize how after 9/11 and for the sake of security and efficiency, ICT platforms and surveillance devices have been expanding all over Europe. The citizenry itself has become an object of digital engineering: movements are tracked, data regarding individuals are stored and matched, and profiles are produced in the search for 'suspicious behaviour'. The citizenry has become an object of technoscientific intervention and an object of distrust. Citizens are suspected by default. Nevertheless, this section suggests that it is difficult to perceive what the real 'threat' is that justifies the deployment of such technological machinery. Paradoxically, by deploying technological means to 'secure' us, the state becomes the primary threat to citizenship. In Chap. 7, Katrin Laas-Mikko and Margit Sutrop address such problems through ethical analysis and discussion. This chapter argues that although public security is touted as the preeminent value, foundational values of liberal democracies such as privacy and autonomy are being violated. These arguments are grounded in the analysis of the implementation of 'second-generation biometrics' in Europe. Software-based technologies and digital platforms are used with the ultimate goal of anticipating actions, particularly malign actions. According to Kristrún Gunnarsdottir in Chap. 6, these are market-driven technologies tailored to recreate people's intentions and anticipate their choices. With a focus on mobility, security and markets, this chapter provides a detailed sociological account of how the ideal of European political integration and social cohesion has led to a reality in which citizens' movements are increasingly under surveillance. The EU integration has first and foremost pursued 'market integration' (and liberalization), which required the free movement of goods, labour force and capital. The regulation of people's mobility, particularly of non-European citizens, is key for this purpose. This chapter also provides insights into how ICT industry is positioning itself as a key actor in the governing of biometric technologies and the production of 'knowledge' regarding citizens. ICT experts have a central role in these developments; however, these experts generally do not appear fully aware of the ethical and political implications of the implementation of this emerging technological field. This chapter concludes that there is no evidence that the implementation of these technologies of surveillance is helping to keep us 'safe', although there is evidence of the market opportunities and benefits that these technological developments are enabling.

In Chap. 7, Kjetil Rommetveit echoes and emphasizes these arguments. By providing a comprehensive analysis of key policy documents, the chapter provides a chronological overview of how ideals of integration have been realized in the EU, increasingly rendering Europe a 'security envelope'. The control of borders is crucial to these endeavours. Two types of imagined technologies are intrinsic to these developments: the imaginary of the 'body as information' and the imaginary of 'systems interoperability'. Each imaginary has entered into the making of a comprehensive legal framework, and combined, they project a certain 'biometrics vision' of European societies. First, information regarding the physical features of citizens is digitalized. Digital identities are contained in identity documents (such as the European biometrics passport). Such travel documents allow tracking the movements of citizens at any time. Second, data regarding people's identities and movements are stored in interoperable systems across European borders. The 'biometrics vision' is realized on two complementary levels: 'promoting the tightened relation between the body and travel documents' and 'the tightened integration of systems and government agencies as given through the concept of interoperability' (p. 118). As this chapter indicates, several problems with data protection, privacy and fundamental rights as well as with the social exclusion of vulnerable groups pertain to the development of the security envelope. The primary argument of this chapter, however, is that these problems are components of a broader set of problems likely to be encountered within large-scale information systems as those systems enter the real world and that the biometrics visions do not truly consider these issues.

If the previous sections of the book have focused on the transformations of citizens' bodies and mobility as those transformations occur in technoscientific governing and governance, the final section introduces a focus on how space is being redefined by technological interventions. This section addresses the emergence of two geo-technologies: geo-engineering and digital maps as developed within GIS (Geographical Information Systems). Both sets of technologies reconfigure orderings and scales. For instance, the implementation of geo-engineering technologies is justified as a way to tackle the global problem of climate change by acting locally (by, for example, manipulating air particles in a particular spot), geo-engineering turns the Earth itself into a technoscientific object whose transformations entail changes on a global scale. With this global scope, geo-engineering is often taken for granted as a field in which citizens' expertise may not have much to contribute. Rather, citizens are to be included as stakeholders. Digital maps, on the other hand, are becoming mundane objects of people's everyday lives. As they operate in mobile phones, iPads GPSs, etc., citizens' lives are becoming increasingly geo-referenced. Citizens are often included as users, co-designers or co-producers in the development of geo-referenced data, software development, digital interfaces and visualizations. However, as the first chapter of this section argues, it is crucial to question how citizens are being included in processes of technological innovation.

Particularly, in the context of digital maps, what type of access to the technology are citizens truly given? What types of agencies are in play? And how might the inclusion of citizens in these processes ultimately empower the people? Because both geo-engineering and digital maps are technologies that have the potential to transform people's lives (in both dramatic and mundane manners), it is important to critically question the working dynamics of the expert-lay interplay. All chapters in this section do so by addressing different aspects of these dynamics. In Chap. 9, Kjetil Rommetveit, Ângela Pereira and Tiago Pedrosa explore how people imagine themselves, others and space through digital maps. This chapter addresses the everyday 'use' of GIS by attending to three types of action: overlaying, interpreting and interacting. The crucial question is whether GIS technologies allow citizens to take an active role within those modalities of action. Descriptions emphasize how 'overlaying' became a prominent tool and skill for depicting and imagining space in the digital age. Overlaying comprises combining several thematic maps (layers) into a single input. Overlaying reveals new sets of relations, data matches and meanings. However, the chapter argues, overlaying is far from a naïve and neutral manner of redefining space (and the relations that inhabit and configure it). Rather, overlaying is informed by interests and values. By the same token, one can ask what type of 'interactivity' digital technologies and, in particular, GIS is enabling. If, for example, the contents of the maps can only be modified and updated by the technology's provider as users provide data (as with Google Earth), the 'interactivity' has ultimately a one-way direction. Alternatively, open source initiatives may allow people to modify not only the content of those maps but, more radically, also the way in which the technology works as well as its use. Thus, the question is what type of new connections these technologies are enabling and the extent to which the technologies enable new forms of collective action and possibly citizenship. For example, private companies may be using GIS technologies to produce geo-referenced data on consumer choices, producing geo-referenced products that are tailored to individual preferences and producing certain social orders by classifying people in certain manners (as on the basis of who lives where). However, new communities of experts are emerging such as 'communities of volunteer virtual cartographers' (p. 136) with specific manners of overlaying and of configuring the relations between the public and the private. It is experience and concern, this chapter argues, that make people care for and create common spaces. Maps created by amateur geographers may not stress accuracy and precision; however, those maps may reflect other concerns based on life experiences, often shared life experiences. The new spaces (and spacing) that GIS technologies afford can mediate new forms of common spaces and collective action as well as privatization and individualization.

In Chap. 10, Fanny Verrax focuses on experts and expertise. The chapter questions the ethical adequacy of GIS professional ethics by analysing a code of conduct that was created by a primary association of GIS professionals. A question is whether general issues of ICT ethics such as privacy are sufficient to ethically address and regulate GIS or whether the technological and social complexity of the technology requires an ethical approach that takes specific issues into account. The chapter suggests that specific ethical frameworks are required that will evaluate issues that emerge with the technology itself, such as the social and political implications of producing massive amounts of digital data on individuals. The expert-lay divide, the chapter claims, should be redefined in accordance with such issues.

In the final chapter of the book (Chap. 11), Paula de Curvelo and Ângela Pereira provide an overview of scientific controversies on geo-engineering. Geo-engineering encompasses a number of technologies and techniques to modify the climate in desired manners. Geo-engineering has emerged (and is politically justified) in relation to climate change, often presented as a technological solution to the non-intended problems that come with highly uncertain changes in atmospheric conditions. It is this technoscientific ambition to control uncertain entities and modify them at will that renders geo-engineering both promising and controversial. Controversy surrounding geo-engineering, this chapter argues, ultimately refers to a rather old issue, namely, 'What is our place in nature?' As with climate change, the dissenters are, on the one hand, those who believe that humans have always modified the climate and that geo-engineering is nothing substantially new or different and, on the other hand, are those who believe that geo-engineering entails a technological manner of intervening in nature with the potential to produce disastrous natural disorders (the proposed solutions to climate change producing more natural and social uncertainty). This disagreement is partially based on the heterogeneous nature of the technology itself. From reforestation for $CO_2$ capturing to the use of chemicals for whitening the clouds to reduce Earth's absorption of solar radiation, a great variety of methods and practices fall within the purview of geo-engineering. Nevertheless, different practices entail different controlling strategies and uncertainties and different implications for governance. Turning the focus towards governance, the chapter identifies three primary narratives informing current debates on geo-engineering. The paper concludes by arguing that debates on geo-engineering can be situated within dominant narratives on science, technology and society such as the set of justifications put forward in European policy strategies that propose technological innovation as the key element for facing challenges, solving problems and triggering social change.

In different manners and in different technological domains, the chapters of this book address changes and transformations that are occurring within science, governance and citizenship. Furthermore, they show how 'transformation' itself is emerging as the object of technoscientific research: the Earth, the human body and mind, and the movement of citizens are becoming objects of design and technological intervention. Finally, 'transformation' is the ultimate concern of technoscientific governance. In the discourse of innovation, social change appears as something that is desirable and can be intentionally shaped by technological intervention. A certain sense of urgency emanates from this narrative, a sense that something must be changed quickly to avoid imminent catastrophe in terms of, for example, environmental disasters, terrorist attacks and demographic collapse. However, the institutional narrative of innovation magically turns problems into challenges to produce a feeling that these are *also* times of opportunity and excitement. Politics drives science, and science drives politics, in novel manners. As presented by institutions, the need for change comes with the imperative that 'we', as European citizens, must take an active role in ensuring our future. Moving towards a sort of 'technological citizenship', citizens are required to be responsible and active in innovation. However, what does the strong desire for change and transformation say about the state of Europe and European politics? What does the desire for change say about European civilization's ambitions? That technology and innovation will enable Europe to be a leading power in a globalized world, as European authorities hope, remains a vision. Providing some conceptual and empirical materials to start situating such civilizatory ambitions and hopes has been the main concern of this book.

Centre for the Study of the Sciences
and the Humanities (SVT)
University of Bergen, Bergen, Norway
ana.delgado@svt.uib.no

Ana Delgado

## References

Bensaude-Vincent, Bernadette, Sacha Loeve, Alfred Nordmann, and Astrid Schwarz. 2011. Matters of interest: The objects of research in science and technoscience. *Journal for General Philosophy of Science* 42(2): 365–383.

Carrier, Martin, and Alfred Nordmann A (eds.). 2010. *Science in the context of application*, 467–483. Dordrecht Heidelberg: Springer.

Delgado, A. 2016. Assembling Desires: Synthetic Biology and the wish to act at a distant time. *Environment and Planning D* (online first).

Delgado Ana, and Roger Strand. 2009. Looking north and south: Ideals and realities of inclusive environmental governance. *Geoforum* 41(1): 144–153.

Delgado, Ana, Kamilla Kjølberg-Lein, and Fern Wickson. 2011 Public engagement coming of age: From theory to practice in STS encounters with nanotechnology. *Public Understanding of Science* 20(6): 829–845.

Dewey, John. 1957 (1927). *The public and its problems*, 3rd ed. New York: Swallow Press.

European Commission (EC). 2001. European governance – A white paper, Brussels. http://europa.eu/legislation_summaries/institutional_affairs/decisionmaking_process/l10109_en.htm. Accessed 25 July 2001.

Frankenfeld, Philip J. 1992. Technological citizenship: A normative framework for risk studies. *Science, Technology and Human Values* 17(4): 459–484.

Guston, David. 2014. Understanding anticipatory governance. *Social Studies of Science* 44(2): 218–242.

Haraway, Donna. 1991. *Simians, cyborgs and women: The reinvention of nature*. New York and London: Routledge.

Irwin, Alan. 2006. The politics of talk: Coming to terms with the "new" scientific governance. *Social Studies of Science* 36(2): 299–320

Joint Institute for Innovation Policy (JIIP). 2012. Study to assist the European research area board: Investing in research and innovation for grand challenges. European Commission, Directorate General for Research and Innovation. http://ec.europa.eu/research/erab/pdf/erab-study-grand-challanages-2012_en.pdf

Latour, Bruno. 1987. *Science in action: How to follow scientists and engineers through society*. Cambridge, MA: Harvard University Press.

Latour, Bruno. 1999. *Pandora's hope: Essays on the reality of science studies*. Cambridge, MA: Harvard University Press.

Lezaun, Javier, and Linda Soneryd. 2007. Consulting citizens: Technologies of elicitation and the mobility of publics. *Public Understanding of Science* 16(3): 279–297.

Nordmann, A. 2007. Collapse of distance: Epistemic strategies of science and technoscience. *Danish Yearbook of Philosophy* 41: 7–34.

Nordmann, Alfred. 2010. A forensics of wishing: Technology assessment in the age of technoscience. *Poesis and Praxis* 7(1): 5–15.

Owen, Richard, Phil Macnghten, and Jack Stilgoe. 2012. Responsible research and innovation: From science in society to science for society, with society. *Science and Public Policy* 39(6): 751–760.

Rheinberger, Hans-Jörg. 1997. *Toward a history of epistemic things*. Stanford: Stanford University Press.

Rommetveit, Kjetil, and B. Wynne. Imagining public issues in the technosciences. Editorial. *Public understanding of science* (forthcoming).

Shapin, S., and S. Schaffer. 1985. *Leviathan and the Air-pump: Hobbes, Boyle and the experimental life*. Princeton: Princeton University Press.

Strand, Roger. 2002. Complexity, ideology and governance. *Emergence* 4: 1–2.

Wynne, Brian. 2006. Public engagement as a means of restoring public trust in science – Hitting the notes, but missing the music? *Community Genetics* 9(3): 211–220.

Wynne, Brian. 2007. Public participation in science and technology: Performing an obscuring and political-conceptual category mistake. *East Asian Science, Technology and Society: An International Journal* 1(1): 99–110.

# Part I
# The Governance of Emerging Technologies

# Chapter 1
# Democracy, Ethics and the Governance of Emerging Science and Technology

**Roger Strand and Silvio Funtowicz**

**Abstract** The governance of emerging science and technologies is becoming an important policy issue. We elucidate the meanings of emerging science and technologies from a multidisciplinary perspective, and we discuss the ethical and political consequences of their introduction in the context of the European Union. We illustrate the relation between science technologies and economic development with some historical post-war examples, concentrating on the limited outcomes of their deployment in relation to the announced promises. Our analysis concludes that the current narrative on science-based innovation and economic growth and employment must be challenged and devolved to an inclusive democratic process. We share our experience from some experimental approaches to the governance of science and technology designed to capture and unfold a progressive political engagement

**Keywords** Condorcet • Eisenhower • Ethics • Governance • Responsible research and innovation

## 1.1 The Governance of Science and Technology

The purpose of this essay is to explore how we, as humans and societies, may and ought to live and deal with technological development. This will be done from a trans-disciplinary perspective, in order to understand the emerging challenges of novel technologies. We will discuss some of these "emerging challenges" as we believe they are perceived by societal actors and institutions, such as the European Union.

A specific scheme of the Seventh Framework Programme (FP7; 2007–2013) of the European Commission (EC) called for the development of "ethical frameworks of new and emerging fields of science and technology". The phrasing is ambiguous,

R. Strand (✉) • S. Funtowicz
Centre for the Study of the Sciences and the Humanities, University of Bergen,
P.O. Box 7805, N-5020 Bergen, Norway
e-mail: roger.strand@uib.no; silvio.funtowicz@uib.no

 
A. Delgado (ed.), *Technoscience and Citizenship: Ethics and Governance in the Digital Society*, The International Library of Ethics, Law and Technology 17,
DOI 10.1007/978-3-319-32414-2_1

as it is not immediately clear which "fields" are being referred to, what is meant by "ethical frameworks", why research funding calls are put forth for developing such frameworks, or why researchers should find it attractive to respond to such funding calls.

In this chapter we will sketch some possible answers to these questions and outline an understanding of the current challenges to governance from a critical perspective to mainstream research and innovation policy. We will outline how both questions and answers are complex, as they are enmeshed into current intellectual and political struggles over language and meanings, as well as visions and ideological perspectives of the present and the future. As an illustration, we start by asking a seemingly innocent question: What is meant by "new and emerging fields of science and technology"? Many would answer that this is related to nanotechnology and the developments of, for instance, new types of nano-structured materials to be used in industrial applications. Others would evoke images of robotics or geo-engineering. We might usefully recall that after the Second World War, the new and emerging fields of science and technology were computing, nuclear power and space travel.

This puzzling expression – "new and emerging fields of science and technology" – was introduced by the EC in its Sixth Framework Programme (FP6; 2002–2006). At that time, the EC mentioned examples such as synthetic biology, which entailed a scientific attempt at designing new living forms. Among the more curious ideas that came up was neuro-economics; the attempt to apply quantitative natural science in order to understand the neurological basis of consumer behaviour. However, a number of quite ordinary-sounding scientific fields have also been called new and emerging, such as applied mathematics, computational science, materials science, and so on and so forth. Moreover, behind a new name such as nanotechnology there is sometimes a quite old and well-known activity: Chemistry, physics, molecular biology or some combination of the three. What is "new and emerging" about it, is not merely a scientific, methodological, technological or temporal issue.

We can sum up the above considerations as follows: First, "emerging science and technology" is not a straight-forward expression. To understand what it means, one has to know and be aware of the context in which the expression was introduced, how it is being used, and how it is transformed as priorities and fashions change. In short, one has to know something about the cultural, institutional and political environment in which this expression is embedded. Second, the illustrations we have presented enable us to assume some of the features of this background: On the one hand, there is an underlying belief in *human progress* through the accumulation of scientific knowledge, the deployment of technological applications, and the hope (and promise) that science-based innovation will contribute decisively to the achievement of political visions and goals. Above all *economic growth* and competitiveness. This is now enshrined in the EU's 2020 strategy, which states that the "Innovation Union" (as the European Union is nicknamed) is essential for a "smart, sustainable and inclusive growth".[1] On the other hand, there is increased awareness, based on recent experience, that science-based innovation is not always welcomed and that it may be actively rejected by citizens and organizations (one example being genetically modified crops or fracking). The costs associated with the promotion and deployment of such innovations is sometimes described and managed in terms of "risks" or "ethical" concerns. Both could be well illustrated by simple questions, such as: Is it a good idea to synthesize new forms of life? Is it desirable to be able to use neurobiology to predict consumer behaviour?

Indeed, what we might call the politics of emerging science and technology is complex, meaning that there is not only a multiplicity of actors and processes but also of legitimate perspectives on the issue. Conventional power- and interest-based explanations are of limited value in this field for at least two reasons. Firstly, the perception of the ambiguous character of some of the developments – the thrill and the terror, as it were – is shared by many actors, also within society, science and technology. Secondly, and importantly, what we are dealing with here are to a large extent things that only exist as hypotheticals or, in the more precise words of Jasanoff and Kim (2009), as sociotechnical imaginaries: collective visions of desirable scientific, technological and social orders.

Unlike many other fields, the actors' interests and passions are not to be taken for granted as something solid and explanatory. Rather, the interests themselves are also in the making and in need of analysis and explanation. The objects of the politics of emerging science and technology are akin to those of magic: They are visions based on imagination, on the way to become reality for some. We mean this in a very literal sense and not as a vivid metaphor. For instance, the EC defined the overall objective of its FP6 NEST (New and Emerging Science and Technology) research funding programme as "improving European anticipation of future scientific and technological needs".

Adding to the complexity, the politics of emerging science and technology has increasingly been framed as an issue of "governance", a concept that also comes with a multitude of meanings. The concept entered the Euro-language quite recently, as a result of what was perceived as a crisis of trust, ironically generated by failures in the emerging science and technology of the time.

From the perspective of authoritative institutions, including Academia, governance has emerged as an essential part of government. We can define governance as "the sum of the many ways individuals and institutions, public and private, manage their common affairs. It is a continuing process through which conflicting or diverse interests may be accommodated and co-operative action may be taken. It includes formal institutions and regimes empowered to enforce compliance, as well as informal arrangements that people and institutions either have agreed to or perceive to be in their interest." (Our Global Neighbourhood 1995)

---

[1] The term "Innovation Union" is now rapidly adopted in EU politico-administrative discourse, both as a concept for its S&T policy, but also in a broader sense, as a new defining principle of the identity of the European Union. See the official EU website for the Innovation Union http://ec.europa.eu/research/innovation-union/index_en.cfm

In the EU context, the word can carry a political or perhaps ethical programme of "broadening and improving government" by more public participation and a larger degree of accountability and transparency, enabling informed participation by the citizens. The undercurrents of concerns represented by "ethics" and "risks" can be detected in the many ambiguities of governance narratives at all levels. In the European Commission's (EC) White Paper on Governance, the despair is visible already in the opening sentence:

> Today, political leaders throughout Europe are facing a real paradox. On the one hand, Europeans want them to find solutions to the major problems confronting our societies. On the other hand, people increasingly distrust institutions and politics or are simply not interested in them (CEC COM 2001: 3).

We interpret this statement as a complaint about a spoiled and ungrateful citizenry. If they (the citizens) could only have been more knowledgeable (in the virtues of science and technology), more interested (in spite of the growing alienation of government institutions) and more trusting (in the power of expertise)! Some (or even many) might disagree with our interpretation. Indeed, this disagreement is a central one in the governance of emerging science and technology.

The issue is not only what kind of scientific and technological processes and products should be allowed, encouraged and publicly or privately funded. It is also a question of the visions, hopes and aspirations for the future development of our societies and civilization, and of fairness, and how power is allocated, distributed and exerted in these societies (Jasanoff and Kim 2009). In von Schomberg's (2011) words, it is a question of "the (ethical) acceptability, sustainability and societal desirability of the innovation process". Some might argue that all this is at stake in all politics. As scholars who do research on science and technology we would respond that science and technology make up a sector of society that holds a special potential for change; a change which is not only of Nature but also of us. The potential is particularly large when the speed of science-based innovation is not matched by the speed of institutions' or people's capacity to cope with the change. The result is a particularly poorly governed (as government and governance) process of innovation possessing characteristic difficulties as a fast-moving, unpredictable, uncontrollable and sometimes invisible target. Invisible also because it has often been surrounded by an ideology in which technological progress is always intrinsically good and which at the same time follows the predetermined path set by Nature as *she* gradually reveals her truths to Science.

On the optimistic side, one can also see the complexities of the coproduction of science, technology and society as a reservoir for the development of new political engagement. Whatever radical or progressive political engagement could mean in our time, we believe it must include and is served by including a critical perspective upon current scientific and technological innovation.

It would take a whole book rather than a chapter to substantiate the claims of previous paragraphs. We will now move on to a few snapshots from the context (historical, scientific-technological, cultural, institutional, economic, and political), to indicate some issues and developments showing the path-dependent trajectory from modern to industrial science, from curiosity-generated science to entrepreneurial techno-science and innovation, and from the creation of public knowledge to the production of corporate know-how.

## 1.2 Historical Snapshots on the Path to the Innovation Union[2]

The classic formulation of the usefulness and benefits of science is usually ascribed to Francis Bacon (1561–1626), who in *The New Atlantis* describes a Utopia of wealth and happiness based on scientific advancements:

> We have also engine-houses, where are prepared engines and instruments for all sorts of motions. There we imitate and practise to make swifter motions than any you have, either out of your muskets or any engine that you have; and to make them and multiply them more easily, and with smaller force, by wheels and other means: and to make them stronger, and more violent than yours are; exceeding your greatest cannons and basilisks. (Bacon 1628/1996, 485–486)

His unfinished manuscript ended with a visionary list of "wonders of nature, in particular with respect to human use". Here are a few examples:

*The prolongation of life.*
*The retardation of age.*
*The curing of diseases counted incurable.*
*The altering of complexions, and fatness and leanness.*
*Versions of bodies into other bodies.*
*Making of new species.*
*Instruments of destruction, as of war and poison.*
*Drawing of new foods out of substances not now in use.*
*Deception of the senses.*

Bacon anticipated that all this could be achieved by the use of the "new tool" of experimental and inductive science. Useful knowledge for Bacon is knowledge about cause-effect relationships that allows us to avoid or bring about the causes of what harms and benefits us, respectively (*Novum Organum* 1620). Knowledge is always better than ignorance because it provides us with more power, that is, more possibilities to avoid harm and achieve good. This argument has been reiterated in science education and science policy ever since. Still, the argument is simplistic in at least two respects. First, Bacon presupposes a consensual "we" who decides what is beneficial and what is harmful: "we" can for instance acquire "instruments of destruction" to achieve more power (over our enemies) and in that way increase our wealth and happiness.

---

[2]This section closely follows the exposition in Rommetveit, Fjelland. Funtowicz and Strand 2013; *"What can history teach us about the prospects of a European Research Area?"* (pp. 38–41, pp. 53–55), from which several of the passages below have been taken.

Secondly, Bacon's examples are of one cause that leads to one effect. In real-world cases, however, there is always only partial knowledge, which is knowledge of *some* causal relationships. When such knowledge is applied, there is the possibility of unforeseen and undesirable higher-order effects; indeed one may identify many environmental problems as the result of such unforeseen and undesired effects of technological intervention into the natural world. In practical terms, even if disputes over what is good and what is harmful are resolved, it remains an open question whether a particular piece of knowledge or technology in fact has created more benefit than harm. The empirical nature of that question is largely ignored in the governance of science and technology.

A particularly explicit formulation of Bacon's Utopia was made by the French philosopher, mathematician and political scientist Marie Jean Antoine Nicolas de Caritat (1743–1794), known as Marquis de Condorcet.

Condorcet identifies himself with the intellectual tradition of Bacon, sustaining that increased scientific knowledge will offer an increase in wealth (Strand 2013: 115–6). Condorcet also believed that an increase in wealth would result in increased equality. For him, inequality is the result of tyranny, violence and prejudice. With Science producing more wealth there will be therefore less motivation to exploit others. However, Science will also propagate Enlightenment and Reason, resulting in less prejudice and violence. Specifically, Condorcet sees the free market – "unrestricted commerce" – as one such mechanism for the propagation of Reason and equality. A few years later, Thomas Malthus published his pessimist view on the future of a growing human population on a finite Earth. Condorcet's reply was the opposite to Malthus'

> [...] if we consider, that prior to this period the progress of reason will have walked hand in hand with that of the sciences; that the absurd prejudices of superstition will have ceased to infuse into morality a harshness that corrupts and degrades, instead of purifying and exalting it; that men will then know, that the duties they may be under relative to propagation will consist not in the question of giving *existence* to a greater number of being, but *happiness*; [...] and not the puerile idea of encumbering the earth with useless and wretched mortals. (p. 273)

We may find in Condorcet most of the ingredients in contemporary political thought: increased wealth and welfare through scientific and technological development and a free market; improved politics and morality through education and social, economic and political science; a theory about mature societies that are sustainable because their rational and morally sophisticated citizens value happiness, reduce their number of offspring and therefore consciously arrive at zero population growth. If we add the belief in the miniaturisation of technology, the information society and the resulting dematerialisation of the economy, Condorcet's programme is essentially identical with those of Western governments since the end of WWII.

Moving another 150 years ahead in time, the well-known report *Science: the Endless Frontier* (1945) written by Vannevar Bush echoes Bacon's and Condorcet's beliefs in science as a provider of practical benefit, now draped in the twentieth century so-called linear model of innovation. Although Bush never spells out the model, he insists that the strengthening of basic research is a necessary precondition for the creation of jobs and the improvement of social welfare. With striking candidness, the report combines Bacon's belief in science, Condorcet's belief in the free market and the fact of post-war consumerist society:

> One of our hopes is that after the war there will be full employment. To reach that goal the full creative and productive energies of the American people must be released. To create more jobs we must make new and better and cheaper products. We want plenty of new, vigorous enterprises. But new products and processes are not born full-grown. They are founded on new principles and new conceptions which in turn result from basic scientific research. Basic scientific research is scientific capital. Moreover, we cannot any longer depend upon Europe as a major source of this scientific capital. Clearly, more and better scientific research is one essential to the achievement of our goal of full employment. (Bush 1945)

For Bush, economic growth is necessary, and it requires that consumers are constantly enticed by new products. The white spots in the world map being filled in and colonial goods already consumed, the West having been conquered, and Europe in ruins and in need of American aid after the War, the United States have to rely on themselves. And hence the title of the report: as all other frontiers are conquered, science is the one that allows for indefinite progress and expansion:

> It has been basic United States policy that Government should foster the opening of new frontiers. It opened the seas to clipper ships and furnished land for pioneers. Although these frontiers have more or less disappeared, the frontier of science remains. (Bush 1945)

The political implication that Bush drew, was the need for a strong public support of basic research. Indeed, his work was influential for the establishment of the National Science Foundation in 1950.

In 1954, The New York Times (September 17) reports of a speech given the day before by the Chairman of the US Atomic Energy Commission, Lewis Strauss to the National Association of Science Writers. Strauss's message resonates with Bacon's New Atlantis:

> Our children will enjoy in their homes electrical energy too cheap to meter … will travel effortlessly over the seas and under them and through the air with a minimum of danger and at great speeds, and will experience a lifespan far longer than ours, as disease yields and man comes to understand what causes him to age.

The promises of nuclear power are nevertheless ambiguous because its intimate relation with the Bomb. Successive generations of new and emergent technologies had to struggle with the ambiguity: the promises of everything for nothing balanced with the terror of the original sin.

There is no going back and we can trace the growing importance of science and science-based technology in the restructuring of the economy and the extension of free trade. In the European integration process, modernity (represented by coal and steel) gives way, slowly first but securely, to the embryonic Knowledge Economy with its own sets of grand promises: human liberation from physical work, natural resource scarcity conquered by technology and cultural substitution and no more "limits of growth". The European coal-miners and steel-workers, fundamental actors of the original Treaty, can now be affluently obsolesced, as modern industries relocate and new jobs require new skills: handling data, developing algorithms and computer models, and continuous innovation of processes and systems.

Witnessing the recurrent economic, financial and political crises, we might well ask if we missed something in our narrative. Perhaps the new, late Modern, phase of co-evolution of knowledge and power has its own pathologies which could have been anticipated. And it was, in unexpected places.

In 1961, Dwight Eisenhower, President of the US, delivered his Farewell Address. The speech is well-known and widely discussed for his reference to the military-industrial complex:

> [...] we must guard against the acquisition of unwarranted influence, whether sought or unsought, by the military-industrial complex. The potential for the disastrous rise of misplaced power exists and will persist. We must never let the weight of this combination endanger our liberties or democratic processes. We should take nothing for granted.

However, another part of the speech is what we are looking for:

> Akin, and largely responsible for the sweeping changes in our industrial-military posture, has been the technological revolution during recent decades. In this revolution, research has become central; it also becomes more formalized, complex, and costly. A steadily increasing share is conducted for, by, or at the direction of, the Federal government. Today, the solitary inventor, tinkering in his shop, has been overshadowed by task forces of scientists in laboratories and testing fields. In the same fashion, the free university, historically the fountainhead of new ideas and scientific discovery has experienced a revolution in the conduct of research. Partly because of the great costs involved, a government contract becomes, virtually, a substitute for intellectual curiosity. For every old blackboard there are now hundreds of new electronic computers. The prospect of domination of the nation's scholars by Federal employment, project allocations, and the power of money is ever present – and is gravely to be regarded. Yet, in holding scientific research and discovery in respect, as we should, we must also be alert to the equal and opposite danger that public policy could itself become the captive of a scientific-technological elite.

Eisenhower describes in simple terms the transformations in the organization, political economy and methodology of science, which have been the subject of increasing academic research during the last decades, and with the traditional disciplines of history and philosophy of science now becoming hybridized with the growing interest in the social studies of science.

Industrialized science (Ravetz 1971) and fungible Mode 2 science (Gibbons et al. 1994) are two among many ways to describe and prescribe new forms of science. The tension between research (more prestigious) and teaching universities reflects the growing pressure to produce economic value through patents and other intellectual property instruments. The result is the privatization of knowledge and relegating the public knowledge production to the category of endangered species. Eisenhower's warning about "public policy" becoming the captive of a "scientific-technological elite" illustrates the ironic nature of the process; when the entanglement of techno-science with power enters a collision course with the humanistic character of the scientific project.

The changes in our understanding of science and the scientific enterprise are important in order to assess the European policies on research and the institutional arrangements put into place to promote research. However, this endogenous view is not enough. We have also to see the effects on humankind and society, and in particular, the consequences on the process of European institutional self-understanding, aspirations and goals. In this context, there is no doubt that modern science and even industrialized science have now become techno-scientific research and innovation, fully at the service of making Europe the most competitive economic block in the world. Techno-scientific research and innovation is funded and promoted with the explicit aim to create jobs, to build a sustainable growth society and to improve the quality of life of the European consumer-citizen in the global market. The operational mechanisms are to remove the obstacles to innovation, to blur the boundaries of the public and private, of business and Academia, to adapt education and training in order to fulfil the goal of making Europe a world-class science performer.

In the 50 years that followed Eisenhower's speech, the process of transformation of science has accelerated, together with the process of economic restructuring, to the point where innovation in the knowledge economy seems to be the only hope of unfettered growth.

What if the reliance on techno-science, co-evolving with emerging structures of power, has put us in a path-dependent *cul de sac*, where to do more of the same, but faster and more efficiently, becomes the only problem-solving strategy possible, increasing the democratic deficit and neo-authoritarian temptations? Eisenhower has a recommendation that we might well consider:

> We must never let the weight of this combination endanger our liberties or democratic processes. We should take nothing for granted. Only an alert and knowledgeable citizenry can compel the proper meshing of the huge industrial and military machinery of defense with our peaceful methods and goals, so that security and liberty may prosper together.

The elements of the "combination" and the relevant actors involved might be different (are they?) but the advice is still valid, only an alert, knowledgeable and committed European citizenry can govern the transcendental processes of innovation taking place; in order to make them consistent with the Humanistic values of tolerance, humbleness, and universality in specificity.

## 1.3 Contemporary Challenges

Still the transformation of science did not occur in a vacuum; societies were also on the move and changing accordingly. With Rachel Carson's 1962 *Silent Spring,* the Western world discovered the Environment and the downside of technological progress, beginning a process that took us to the 1992 United Nations Conference on the Environment in Rio de Janeiro. The under-currents of concern and awareness grew slowly but steadily, with risks and ethics now publicly discussed and somehow channelled through emerging narratives of participation, precaution and sustainability.

Other important aspects of this process are worth citing. The first relates to the discovery that only occasionally the hopes and expectations justifying new and emerging fields of science and technology were realised, and that this kind of progress not only had a downside but that it was even difficult to grasp its upside. The second dimension refers to accumulation of failures and disasters (De Marchi 2015), putting a question mark on the wisdom of relying blindly in an expertise that promises everything at the cost of nothing. Controversial new and emerging science and technology in the fields of food and health add a new focus of concern, and being so close to human identity and life-styles, generate new questions about risks and ethics.

Finally, there has been an expansion of the world media and the Internet, enabling real-time and decentralised communication. We should not underestimate the power of images showing the effects of accidents and disasters, and of experts disagreeing publicly. Esoteric debates are now ventilated in the open, and the citizens can see that, firstly, science does not speak with one voice, and secondly, that scientists are not in control of the artefacts of their creation. Nuclear power, from Three Mile Island to Fukushima, is the paradigmatic example.

We argued about the ambiguity (and complexity) of expressions like "emerging fields of science and technology" and "ethical frameworks" and we mentioned the under-currents of concern expressed in terms of risks or ethics which were mainly devolved to science (for risk assessment) or to bureaucracy (for ethical checklists and committees). This mechanism seems not to work anymore, and this is especially true now, when the transition from modern science to goal-oriented interest-based techno-science is almost complete, and this is perceived and acknowledged and contested by a large part of the citizenry. The search for new frameworks – whether we call them ethical, political or of governance – therefore needs to be free to consider new pathways that depart from the usual institutional arrangements. This is the answer to the initial question of this chapter: why researchers should find it attractive to respond to funding calls for the development of "ethical frameworks of new and emerging fields of science and technology".

The methodological implications of such a vision are in themselves radical. First, there can be no absolute distinction between research and the political processes in which the research is undertaken. In more general words; between knowledge and action. As researchers we are engaged; any other idea is illusory, and even undesirable. Secondly, this means that there is no absolute distinction between researcher and citizen. Finally, above we have indicated how attention to narratives and imaginaries is required to understand the development of emerging science and technology as well as its governance. Indeed, the Innovation Union *is* an attempt at creating a powerful imaginary. This importance and the primacy of imaginaries and narratives over ethical argument and risk calculus is recognised, we believe, everywhere else in society than in the particular science- and ethics-based institutions of governance. Their deficiency in this sense seems to be closely related to their being dysfunctional with respect to the under-currents of concern. Being both scholars and engaged citizens in the co-construction of alternative pathways for governance, however, our understanding is that we need to engage in the creation of imaginaries and narratives, and not only study them.

As might be well known to many readers of this book, there are many examples of experimental projects and approaches that intend (or purport to intend) to open up the governance of emerging science and technology to broader issues and concerns; democratize it by bringing other stakeholders and citizens into processes of agenda-setting and "engagement"; and acknowledge and work better with aspects of imagination and the potential for narratives and creativity. Furthermore, many of these efforts have been taking place within or with funding or other endorsement by public research and research-funding institutions, under umbrellas such as "public participation", "public engagement", "science and/in/with-and-for Society" (EU), "anticipatory governance" (the US) and lately "responsible research and innovation" (EU and several European countries). Indeed, it is often in these contexts that the term "governance" is explicitly used in R&I policy; in the close to hegemonic processes of business-as-usual policies of "innovation for growth" the term governance is never central if used at all.

The term "hegemonic" is of course value-laden; it entails the evaluation by which something is considered to dominate. It entails the evaluation of the experimental efforts as marginal. In economic and administrative terms they indeed are. The securing of the Science-with-and-for-Society programme within the eight framework programme of the EU (Horizon 2020) is noted as a remarkable achievement; it holds 0.6 % of the total budget of Horizon 2020, which again disposes of a few percent even of *public* research funds within the European Research Area.

Still, many of the authors of this book have devoted considerable energy and engagement into these experimental projects. For instance, the authors of this chapter worked within the FP7 Science-in-Society projects TECHNOLIFE and EPINET.[3] TECHNOLIFE was a response to the EU call already mentioned above, on the need for "new ethical frameworks" for emerging science and technology. EPINET, in a similar fashion, responded to an EU call on the need for new frameworks for integrating different methodologies for assessing (non-monetary) societal impacts of emerging science and technology.

Like in many other such research projects, TECHNOLIFE and EPINET resulted in a lot of activity that made it possible for us as project coordinators to conclude with success by our own criteria. We developed methods and published results. In TECHNOLIFE, we created, explored and tested new internet-based tools for citizens and other stakeholders to share and discuss sociotechnical imaginaries and ways of analysing such discussions that could enrich the scope of ethics debates within the institutions (ethical reviews etc). In EPINET, we created, explored and tested new ways of bringing S&T researchers and innovators in close contact and dialogue with ethicists, STS scholars and other types of technology "assessors". Although such claims are hard to document, we think we are justified to believe that many of the people involved in the two projects broadened their vision of governance of science and technology and were left with an enriching learning experience.

---

[3] Documentation, working papers and open access publications can be found at http://www.technolife.no and http://www.epinet.no

Still, the impact on the overall governance of science and technology of these two or indeed the whole suite of such projects is hardly visible from the bird's eye perspective. In any case we do not believe that strong political engagement is motivated and initiated by calculation of the expected outcome. Rather, it may be part of the problem of technological hubris to think that a "good" governance of science and technology, be it in terms of sustainability, social justice, the humanistic tradition or other fundamental rights and values, is something that can be obtained by clever social engineering (Nordmann 2010). Accordingly, we have previously argued for a motivation of such experimental projects and endeavours as a question of *commitment* (Funtowicz and Strand 2011; Strand 2013), a commitment that should not let itself be too easily disappointed by the lack of "hard" evidence as judged by criteria that themselves are part of an instrumentalist type of governance.

We remarked above that the acknowledgement of complexity and uncertainty also may provide hope: Positive developments may take place, contrary to prediction, expectation and analysis. Part of the reflexive role of us as scholars of science, technology and society is therefore also to optimistically note that we may be wrong about the hegemony and pervasiveness of the innovation discourse and the marginality of our own efforts. Another part of the reflexive role is to point towards the needs and opportunities for other actors than ourselves. The political initiative required cannot be ignorant of and in perfect isolation from current institutions, but neither can it be mobilised from within the institutions, of which we are hardly independent (Strand 2013). Our highest hope is therefore that this chapter and this book will be read by ordinary people beyond the usual circles of research and innovation policy-makers and scholars studying them, and that it will inspire them to take up the challenge of governance of science and technology as one of the most important political issues of the twenty-first century.

## References


Bacon, F. 1620/1994. *Novum Organum*; *with other parts of the great instauration*. Chicago: Open Court Publishing.

Bacon, F. 1627. *The New Atlantis*.

Bush, V. 1945. *Science, the endless frontier*. United States Office of Scientific Research and Development, U.S. Govt. Print. Off.

Carson, Rachel. 1962. *Silent spring*. Boston: Houghton Mifflin.

CEC COM. 2001. 428 final (Brussels, 25.7.2001) *European governance* A white paper. http://eur-lex.europa.eu/LexUriServ/site/en/com/2001/com2001_0428en01.pdf.

De Marchi, B. 2015. Risk governance and the integration of scientific and local knowledge, Ch. 9. In *Risk governance: The articulation of hazard, politics and ecology*, ed. U. Fra Paleo, 149–165. Berlin: Springer. ISBN 978-94-017-9328-5.

Eisenhower, D.D. 1961. Farewell address from President Eisenhower. http://www.ourdocuments.gov/doc.php?flash=true&doc=90.

Funtowicz, S., and R. Strand. 2011. Change and commitment: Beyond risk and responsibility. *Journal of Risk Research* 14(8): 995–1003.

Jasanoff, S., and S.-H. Kim. 2009. Containing the atom: Sociotechnical imaginaries and nuclear power in the United States and South Korea. *Minerva* 47(2): 119–146.

Marquis de Condorcet (M.J.A.N. de Caritat). 1796. *Outlines of an historical view of the progress of the human mind*. Philadelphia: Lang & Ustick.

Nordmann, A. 2010. A forensics of wishing: Technology assessment in the age of technoscience. *Poiesis & Praxis* 7(1): 5–15.

Our Global Neighbourhood. 1995. *The report of the commission on global governance*, 2. Oxford University Press.

Ravetz, J. 1971. *Scientific knowledge and its social problems*. Oxford: Clarendon.

Strand, R. 2013. Science, Utopia and the human condition. *Foresight and Innovation Policy* 9:2/3/4.

UN Conference on Environment and Development. (Rio de Janeiro, 1992). http://habitat.igc.org/agenda21/index.html; http://ec.europa.eu/europe2020/index_en.htm, EU 2020 Strategy; http://ec.europa.eu/research/fp6/pdf/fp6-in-brief_en.pdf, 6th Framework Programme (6FP); http://cordis.europa.eu/fp7/home_en.html, 7th Framework Programme (7FP).

von Schomberg, R. 2011. Prospects for technology assessment in a framework of responsible research and innovation. In *Technikfolgen abschätzen lehren: Bildungspotenziale transdisziplinärer Methode*. Wiesbaden: Springer VS.

# Chapter 2
# 'The Public Spectre': A Critical Concept of Public Engagement with Technology

**Kim Sune Jepsen, Ana Delgado, and Thora Margareta Bertilsson**

**Abstract** This paper is concerned with how forms of publics come into being in situations of technological innovation and change. The paper attends to sudden social formations which arise as responses to what is perceived of as threatening socio-technical inventions into the routine of everyday social life. We revisit the Dewey-Lippmann debate, and we look at spontaneous citizens' reactions towards emerging technologies such as Google Street View as well as we discuss some institutionally organized events of public engagement with science. As we explore the formation of publics inside and outside various institutional contexts, we suggest a concept of mobilizing unexpected agencies that we call the public spectre. Crowds emerge spontaneously, and assume the figure of a 'public spectre' that resides in the unforeseen. When repeated as collective events, crowds stabilize and assume the figure of publics. The notion of the public as a spectre draws attention to the plurality of forms in terms of which publics emerge and take form.

**Keywords** Public spectre • Social formations • Governance of science and technology • Radical democracy

K.S. Jepsen (✉)
Department of Sociology, Lund University, Lund, Sweden
e-mail: kim_sune.jepsen@soc.lu.se

A. Delgado
Centre for the Study of the Sciences and the Humanities (SVT), University of Bergen, PO Box 7805, N-5020 Bergen, Norway
e-mail: Ana.delgado@svt.uib.no

T.M. Bertilsson
Department of Sociology, University of Copenhagen, Copenhagen, Denmark
e-mail: mb@soc.ku.dk



A. Delgado (ed.), *Technoscience and Citizenship: Ethics and Governance in the Digital Society*, The International Library of Ethics, Law and Technology 17,
DOI 10.1007/978-3-319-32414-2_2

## 2.1 Confronting Technology: Accommodating to Threatening Events

In 2010, citizens in the city of Berlin opposed the introduction of Google Street View in their city spaces. *Der Spiegel*, a main German weekly, reported that more than 250,000 German citizens asked Google to block the images of their houses. These citizens experienced Google Street View as a technology that allows them to watch and to be watched. 'Privacy' was made an issue of citizen concern and their manifestation of a public confrontation. German government subsequently asked Google to impose restrictions on the use of the application in those cities where public response emerged. Berlin artists also contributed to channelling public response: in 2011 the 'Free Art and Technology' group produced a fake Google Street View car. Using GPS technology they tracked how the car 'conquered' the streets of Berlin, thus, ironically 'watching the watcher'. Similar concerns with privacy have lately been expressed by sectors of the citizenry in a number of European cities regarding the ways in which data about their health, housing or movements is stored. In 2012, the European Commission proposed a revision of the EU Data Protection Directive in Europe in the shape of a Regulation that impose restrictions to technology not found in the US or Asia. This proposal has proved highly problematic not the least because of corporate interest; up to date the new Regulation is still being negotiated. But how do European citizens experience the technological invasion of their privacy?

The three authors of this paper participated in a project called Technolife, in which such developments as to technological invasions of privacy were the focus.[1] Using media materials, the researchers in the project invited affected European citizen's to discuss how they experience technologies such as biometrics and Geographical Imaging Systems (GIS, see Rommetveit et al. this issue). The forum aimed at visibilizing those voices that often fall out of normal representative politics. But as the debate developed, the researchers coordinating the forum had to redefine some of the initial suppositions. For instance, the conversations in the forum revealed that privacy is not necessarily a matter of being 'reserved' or only an attribute of individuals. Rather, in those online debates major concerns were expressed over technologies that would enable the classification and eventually the discrimination of certain groups. A summary of the results from this discussion forum stated how: "What counts as privacy appears notoriously difficult for participants to clarify except in reference to either breach or control" (Gunnasdóttir et al. 2011: 10). In this forum, concerns were raised as to how massive amounts of data on people's bodies, daily behaviours and choices could be stored in centralized ways, and who would eventually have control over that data. Concerns were also raised as to the potential implications of sorting individuals in groups as 'types'. Both corporations and states were seen as sources of distrust. This imaginary of information control and distrust was echoed in the protests against Google Street

[1] Technolife: Transdisciplinary Approach to the Emerging Challenges of Novel Technologies: LifeWorld and Imaginaries in Foresight and Ethics. FP7.

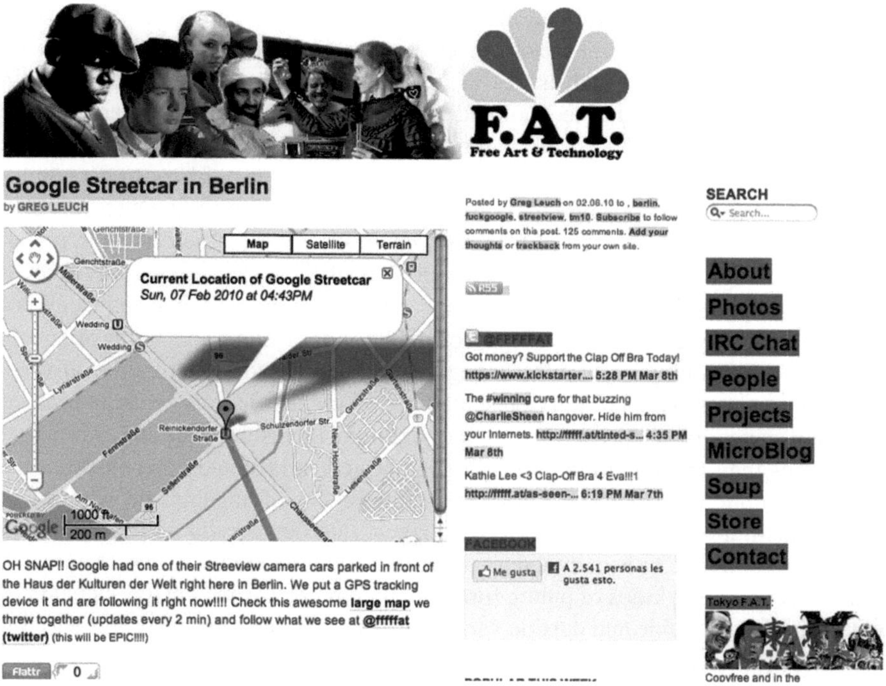


**Image 2.1** Description of the "Google Street Car" project as presented in the website of the "Free Art and Technology" group website. http://fffff.at/google-streetcar-berlin/

view in Berlin. Possibly such imaginary is connected to past stories of technologies of war, spies and information control as experienced by the population of Berlin. Nevertheless, the Technolife debates and, especially, the ways in which issues developed in the course of such debates, suggest that issues and publics simultaneously develop grounded in collective and daily experiences[2] of technological change. Often experienced in a way or another as being 'out of control', technologies disrupt in the everyday life of citizens as disturbing elements that need to be somewhat 'accommodated' (Image 2.1).

[2] We do agree that if 'experience' is to be used as an element to explain the emergence of publics and public issues, it should be historicized rather than too simplistically be used as an explanatory device. As feminist historian Joan Scott argues, 'experience' should not be used as "the origin of our explanation, not the authoritative (because seen or felt) evidence that grounds what is known, but rather that which we seek to explain, that about which knowledge is produced" (Scott 1991: 780). Used as given, 'experience' would prevent us from looking into the conditions that made different experiences possible. As it is, this essay lacks a systematic empirical investigation of citizen's differing experiences with technological development. Furthermore, we focus on particular types of technological systems that are imposed and arrive at people's life as disruption, paying less attention to the liberating and empowering experiences that people can have to certain type of 'open' technologies (see for instance all the new ''DIY' wave in Europe in Nascimento et al. 2014. See also Murphy 2004). All this can be seen as shortcomings. Yet, the aim of the essay is simply to characterise a political figure, the public spectre, that readers can evoke and question when reading the following chapters of this volume.

The story of public confrontations with Street View in Berlin is one of the many stories illustrating how, in an age characterized by rapid technological development, citizens actively engage with issues of accommodating unfamiliar and uncertain implications of science and technologies. As technological innovation and change disrupt and interfere with social communities and shared lifeworlds, citizens are increasingly confronted with technologically invoked issues of how to accommodate to strange and imposing environments. As the story above suggests, citizens felt uneasy as to the prospects of the new technology entering their lives: in rejecting a technology that might threaten their privacy, their citizenship was invoked.

In this paper we explore and conceptualize how emerging science and technologies are currently enabling new forms of public mobilization and engagement. In modern societies citizens have to engage with science and technology in diverse ways: Sometimes, as in the story presented above, citizens identify with an issue of common concern (i.e. privacy) and they organise themselves around this issue; sometimes they are invited to participate in public hearings and other institutional forms of public participation. Across such differences of engagement within and outside institutional policy-settings, we suggest that democratic politics is taking form as a challenge to accommodate unsettling socio-technical issues, eventually, sparking conflicting kinds of public formation into being. As we pursue such democratic challenges inside and outside various institutional settings, we propose a concept of unexpected and contesting agencies that we call the public spectre.

We will briefly present a context of social life, where the representation and estrangement of public life has been subject to debate. We focus on how the role of policy institutions, with their multiple public engagement exercises around different European countries, impinge on the prospects of democratic life. This issue – how science and technologies are institutionalized and governed – has not least been critically explored by science studies of the manner public concerns are often subordinated to expert cultures and epistemic knowledge claims that disavows their normative and political commitments (e.g. Jasanoff 2012; Welsh and Wynne 2013; Wynne 2006). Targeting how people's concern with technological change are restricted by pre-structured frameworks and roles in policy-making, we move towards more constructive analysis of alternative public formations. We reinterpret the notion of phantom public (Lippmann 1925; Marres 2005a) and propose an auxiliary notion of the public spectre: (1) communal concerns and social formations engaged with science and technologies are emerging outside political representation; (2) such alternative formations are significant agencies for articulating and accommodating the social implications of science and technologies. In the following section, we elaborate on these points in order to formulate a widened notion of public politics in high-tech societies.

## 2.2 Techno-politics: Trust, Risk and the Crisis of Representation

In high-tech societies, a problem is the persistent lack of inventive democratic politics capable of accommodating the rapid growth of unfamiliar and uncertain issues of science and technology. STS scholars have thus argued that we live in a society characterized by a dual speed of uncertain techno-politics. This duality is played out between (1) the many unfamiliar socio-technical inventions that disrupt our life-world; and (2) an increasing inability of representative governments to offer durable political accommodation to the issues at stake in people's lives. There seems to be a mismatch between the rapid transformation of social everyday life and the institutional mechanisms of slow representative democracies. In an attempt to deal with this crisis of representation, a 'participatory boom' has emerged within the last decades. However, such 'speeding up politics' through more institutionalized opinion-making, might simply supersede more profound public concerns with technology. To tackle this predicament we will engage with the crisis of representative democracy in relation to the advance of techno-science.

### 2.2.1 *The Crisis of Representative Democracy: Exercising and Experiencing Public Life*

The crisis of representative democracy has recently crystallized in government and policy-regulations that show concern with an apparent lack of public trust in current expert institutions. This was very clearly expressed by the European Commission White Paper on Governance in 2001. This document expresses worries with the growing distance between public interests and the state of scientific expert knowledge. This is portrayed as the main reason for an increasing lack of representation and public distrust (EC 2001: 1, 7, 12, 19).

In an attempt to cope with the crisis of representative democracy in the last years, and especially in Europe, there has been a proliferation of public consultations and institutional events of public engagement with science and technology. As Allan Irwin has called attention to the way science and technology policies sweeping across the political landscape of Europe currently coalesce in a 'frequent talk of dialogue' and democratic inclusion of public concerns: 'blend modernistic assumptions of sound science, institutional control and administrative rationality with a language of two-way dialogue and taking citizen concerns seriously' (Irwin 2006: 304). In this respect, repeated institutional calls for more public engagement is an important democratic gesture that must not least be valued in terms of what kind of actual possibilities for public engagement with science and technology it allows for.[3] Thus, what are

---

[3] This theme is widely addressed. For some recent science studies of the obstacles to public engagement in democracy see e.g. Leach et al (2005).

the political and ethical implications for democratic life, when technological change is addressed inside institutional structures that seek to govern public opinion?

In practice consultation exercises like citizen juries and consensus conferences, enact a procedural approach to technological change in democracy. Representatives of different kinds of expert institutions stand trail to a small select committee of citizens that perform an interrogation and evaluation of relevant technological issues. A public viewpoint can thus emerge and assent may be given to issues that need more careful attention by experts before a consensus. In this sense the political ideal is that democratization of technology occurs through public opinion-making. Although such institutional venues are important for securing democratic engagements, the following example demonstrates how such procedural approaches may also entail unintended public criticism of the democratic representation involved.

In a revealing study (2005) of the Danish Board of Technology and its consensus-building conference on the future of Danish Electronic Patient Recording (EPR),[4] Casper Bruun Jensen experienced how a strictly orchestrated political setting was suddenly reversed. A few 'disinterested citizens' had been selected to deliberate and envision the usefulness of EPR before implementation could commence. Experts had manufactured a catalogue of issues for deliberation. This catalogue was based on the idea that EPR was more efficient and entailed better information acquisition of patient data. Thus, Jensen observed the catalogue rather proved to orchestrate the spectrum of salient issues, making other issues redundant. The issues discussed remained largely stable during the process of citizen's queries; a sign of the way citizens were mainly informed and guided towards certain expert understandings rather than interfering with their own viewpoint.

However, on two accounts citizens managed to voice more critical issues about EPR that also affected the nature of the procedure. Firstly, citizens raised an issue with how to secure better data access seen from a future 'user perspective' rather than the present 'expert perspective' Secondly, at the public presentation of the consensus rapport citizens voiced an inability to represent the target group of the technology: those too weak or ill to mobilize a public voice (Jensen 2005: 228; 232). However, notwithstanding the emergence of such public concerns the procedure was cast in terms of a consensus.

This is an example of how current institutionalized democracy seeks transparency in consensus-building while, paradoxically, re-enforcing structures of unacknowledged normative powers that restrict the spectrum of concerns to be raised.[5] Similar to Jensen's observations the public deliberation on the problems of gene modifying technology in the UK 2006 (*GM Nation?* debate), was modelled on the assumption

[4] The particular conference was modelled after 'development space' that differs somewhat from consensus-conferences. The former is a more open process by being a more open process of deliberation. However, both procedures relay heavily on technical expert knowledge as key force in informing citizen's opinion. The aim is often identically to generate consensus and the writing of a final rapport with political recommendations to the government (on this issue see Blok 2007)

[5] Jensen stresses that consensus conferences are rather to be seen as experimental public encounters e.g. between expert/lay cultures of normative reasoning.

that the 'ordinary citizen' equals the 'disinterested', and 'uninformed' person knowing little about new technologies and related issues to be deliberated (Irwin 2006). Hence, 'partisans', those with a recognizable normative concern, were dismissed from participation in the public deliberation exercise (Lezaun and Soneryd 2007). Yet, the deliberative procedures of EPR and GM Nation? engaged an important political momentum. Although both exercises were consensus-driven, citizens in the EPR managed to voice a concern with lacking representation of the 'absent others'. Likewise, the managers of the *GM nation?* recognized how 'partisan' citizens associated and voiced culturally entranced and often negative concerns with GM technology at local meetings and official websites 'in the wild' (Lezaun and Soneryd 2007). In both cases, the exercises entailed unforeseen public formations that questioned the representative powers at work. How might we conceive of such public formations?

Within the last decades, one can observe an increasing tendency to include citizens in roles as 'users' or 'consumers' in exercises of public engagement with science and technology (i.e. citizens as consumers of new nanotech products). This can be seen in connection with an emphasis on innovation presented as a user-driven process. When the representation of citizens is staged in such reductionist manners, valuable experiences and perspectives of how citizens see themselves affected by technological change are in danger of being ruled out of democratic politics (Wickson et al. 2010). Peoples' perspectives seem to be useful to the extent that they may help to justify expert reasons and wider institutional needs of generating *certain* policy-outcome (i.e. economic growth). At the bottom line, such orchestrated policy settings imply that people investing creative visions can primarily provide the means to raise sales numbers of developing technologies. Under such conditions current talk of 'empowered consumers' becomes a political oxymoron. The talk conceals new forms of technological domination as citizens involved are effectively relegated to the margins of representational political powers. Metaphorically speaking: in the role of the 'valuable stakeholder' or 'consumer' citizens' political concerns with technology easily become significant artillery to be used by different expert institutions in view of their own objectives. The institutionalised 'participatory boom' that we have been experiencing in Europe since the late 1990s is an important attempt to regain public trust.[6] But in practice, participatory exercises are also a way of producing domesticated publics, when parts of the citizenry are selected in accordance with pre-given criteria, while citizens not fitting such criteria are excluded from participation.

We suggest that a critical appraisal is needed of how policy-institutions are orchestrating publics into being and the ways in which *alternative democratic politics may be emerging*. As Sheila Jasanoff has argued democratic society is in need of democratic politics that addresses ethical issues of how people living at the margins of political power are affected by science and technology (Jasanoff 2003, 2005). What we need is new ways of recognizing public concerns in the effort to democratize science and technologies.

---

[6] "On one hand, Europeans want [politicians] to find solutions to the major problems confronting our societies. On the other hand, people increasingly distrust institutions and politics or are simply not interested in them." (EC 2001: 3).

## 2.3 Sparking Publics into Being: The Dynamics of Formation

In this section we address the dynamics of public formation conceptually. We introduce and reinterpret Noortje Marres' notion of the 'phantom public'. Building further upon her interpretation of the enigmatic concept of a public life struggling with technological change outside representational power, we suggest the need of an widening notion that emphasizes the contesting nature of public responses in relation to intruding technology, and that we call the 'the public spectre'. The public spectre is a social figure that reflects concerns with neglected and estranged issues outside diverse orchestrated frames of public participation. Hence, a public spectre marks an instance of communal re-emergence and political actuality in the context of accommodating uncertain implications of science and technologies. 'To offer accommodation' is crucially about a concern with *doing justice* to neglected experience and histories of technologically affected people. Hence, the public spectre refers to the articulation of neglected concerns that may provide important alternative understandings of how science and technologies matter.

### *2.3.1 The Politics of the Phantom Public*

In the following we will discuss in more detail Noortje Marres' notion of the phantom public as it relates to alternative forms of public formations and the notion of public spectre. Marres has thus proposed that publics emerge as an association of many 'political outsiders' around specific challenges of settling issues that remain invisible for governments and expert agencies: '*When issues risk to be deserted by the agencies that should take care of them, the public steps in as the caretaker of these affairs*' (Marres 2005a: 48).[7] Building on a particular interpretation of the Walter Lippmann (1925) and John Dewey's (1927) debate on the nature of publics and the public sphere, she suggests that publics emerge as they get involved with specific issues that the authorities do not know how to handle, or even what these are all about (Marres 2007). As Marres proclaims, the disserted issue '*sparks public involvement in politics*' into being (Marres 2005a: 66).

Citizens that come together in a public formation are not necessarily characterized by commonalities. On the contrary, citizens come together in the face of issues that cannot be handled by a community: '*…what the members of a public share is that they are all affected by a particular affair, but they do not already belong to the same community: This is why they must form a political community, if the issue that affect them is to be dealt with*' (Marres 2005b: 214). Public formations emerge in response to the way technological change disrupts established orders of social life

[7] 'The public must become involved in politics, if its (potentially) harmful effects are to be addressed' (Marres 2005a: 51).

by its. Similar to the notion of sub-politics (Beck 1992), publics emerge as an association of political 'outsiders' that struggle over the right to deal with uncertain consequences of indeterminate technological change that affect their social life differently: *'the emergence of a public affair can be characterised as an occasion on which a specific irreconcilability between modes of living comes to be articulated, as opposed to the many divergences among that are often easy to observe, but rarely translate into focused disagreement'* (Marres 2005a: 59).

The suggestion is thus that democratic politics concerns the articulation of a controversial issues that in turn becomes the object of a public mobilization : *'In the situations in which political forms cannot contain the effects of change, what I call issues appear as an organizing principle of the public'* (Marres 2007: 769). Hence, publics fluidly *appear* and *disappear* depending on how the struggles over issue formation occur in daily life. In a suggestive gesture Marres has called this state of uncertain experimental politics a *phantom public*.[8] Publics appear as inherently uncertain (and heterogeneous) experimental spaces outside and invisible to the control of institutionalized forms of representative powers. As an example Marres argues that European vegetarians and Kansas agricultural business that 'decide to splice a pig's gene into their tomatoes' will properly have completely divergent lifestyles and values but be mutually entangled in an issue of genetically modified (GM) food, thus becoming a public (some might see this as a multiple and confronted public though). In such situations both parties are placed in need of taking into consideration 'what effects it has on others' and reach an agreement in order to successfully carry on their way of life: *'In order for successfully 'GM-ify' agriculture in Kansas some sort of settlement must be found for the resistance of European vegetarians'* (Marres 2005a: 59).

Marres' notion of the phantom public brings about an important recognition of a ghostly feature of current democratic politics: publics appear and disappear as indeterminate and unfamiliar issues interfere with citizens' ordinary life. Phantom publics appear in the presence of issues that are institutionally invisible and politically unsettled. Furthermore, such issues invoke publics that are heterogeneous and socially distributed. And yet, how is the collective invoked? How are issues invoked and articulated? Let us return to the story of the Google car in Berlin: 'privacy' in relation to Street View emerges as a public issue insofar as there is a group of people that is concerned about 'being watched'. There was already an existing concern based on a shared life world and experience (perhaps a constellation of stories of surveillance during the Nazi/Stasi Regimes) that spurred people's reaction to Google Street View as a threat to privacy. In the case of GMOs, arguably, the issue is not one, but it unfolds and is articulated as multiple. Health, genetic contamination,

[8] Marres reinterprets Walter Lippmann's concept of the Phantom Public (1925). Lippmann argues that public opinion only has democratic value when it assent with the most rationally fit representative agency that can govern in the event of extra-ordinary crisis. Any notion of a common political purpose or will to government is an illusion or phantom, because citizens normally remain ignorant 'outsiders' to the complex expert issues and regulations that must guide social order (Lippmann, p. 68, 125). Marres reverts Lippmann's argument and suggests a positive affirmation of politics outside representative powers: a phantom public.

economic growth or farmers' rights and global justice were emphasized by sectors of the public differently revealing a variety of disputed concerns. With this examples in mind, we can reinterpret Marres' notion of the phantom public: Yes there are issues that are misrepresented or simply ignored by institutions; Yes, these may become political phantoms, sparking publics into being; but this is only possible insofar as there are people around who share common concerns, however diffuse and ways of being in the world. Putting other differences aside people made common cause in demonstrating a concern with *rights to representative power* in the face of technological advances. Thus, Google Street View marketed itself with a promise of offering freedom to 'let you explore places around the world'.[9] But a configuration of past experiences with fears and grievances might have re-emerged in a counter-memory that made Google technology into an instant public issue. In our view, publics become 'issue-related' within a context of existential concerns; issues are bound to shared emotions and ways of experiencing the life world (rather than only to a pool of shared individual interests as Marres ultimately suggests).[10]

The notion of the *public*-spectre is a 'social figure' that reflects potentially normative concerns with justice and rights to representative power (dis)entangled in technological change. As our stories suggest the public spectre connotes the communal aspirations re-emerging in citizen's confrontations with technology. Hence, reflecting on Google Cars Street View of the social world and the salience of issues entangled in EPR-technology and the *GM Nation Debate,* we suggest that a common pattern is the public formation around common experience gaining unforeseen political momentum. In the extent to which such communal concerns take form outside the orchestrated frames of representation they emerge as the unexpected and neglected form of a public spectra: sudden social formations emerging from the margins in the form of contestation and eventually capable of setting new political orders.

[9] https://www.google.com/maps/views/home?hl=en&gl=no

[10] At the present, in the Summer of 2013, another 'public spectre' is potentially forming at diverse places in the world, in Brazil, Germany, Scandinavia and Britain, with the revelation that US run agencies are in possession of worldwide surveillance programs. A 'public spectre' threatens to erupt against uncontrolled US power, a prospect that apparently worries also President Obama. A worldwide petition is now circulating suggesting that Eric Snowden, now a refugee in Russia, receives the Peace Nobel Prize so as to balance the previous Nobel Prize given to President Obama 2009. A 'public spectre' needs not always be physical, but can as well be mediated by texts and images triggering the concerns and emotions of people who are physically apart.

## 2.4 The Notion of the Public Spectre: Towards Pragmatic Grounds

Marres' proposition of the phantom public is important for understanding the potential antagonistic nature of democratic politics, the role of non-human agencies and uncertain issues for democratic mobilizations.[11] Yet, we propose that an alternative strand of public engagements with science and technologies revolve around the manner in which people with shared or else diverse common concerns are affected by uncertainties of science and technologies but which are neglected in the current forms of political representation. What the public spectre comprises is thus the possibility of genuine democratic rupture through which dominant forms of representation are contested and reclaimed on the ground of the subordination of marginal voices and 'othering'.[12] Crucial to such formation is thus also persistent social issues of *rights to representative power in voicing collective concerns.*

Thus, in the case of EPR-technology we saw how a public concern with issues of social justice and representative capabilities, paradoxically, emerged through a double loss of representative power. The inabilities of participants to represent echoed the *compassion for* those who were marginal and outside any political representation; those made absent but embodying a valuable experience as the possible 'future users'. In the *GM Nation?* we saw how politically motivated citizens mobilized outside institutional settings to voice unforeseen resistance to the representative politics of GMOs. People in Europe aligned with peasants in Latin America and other parts of the world confronting Monsanto while articulating the issue of GMO through claims for health, sustainability and justice. In the case of Technolife, participants voiced unforeseen concerns with the social control by the state and corporations and in Berlin, Google Street View become object to a social formation of *commonly* concerned citizens. In all examples, it was arguably a communal concern that suddenly gained political momentum and changed the course of events: a new affiliation of political identities emerged by also offering accommodation to uncertain technological issues of science and technologies, thus, also enacting a common need for a widening technological citizenship. In the extent to which such communal concerns take form outside the orchestrated frames of representation they emerge as the unexpected and neglected form of public spectra, sudden social formations eventually capable of uprooting the present and setting new political orders.

---

[11] On the subject of how the phantom public is concerned with settling controversial issues and non-human agencies see also Bruno Latour's notion of Ding-politik (2005).

[12] The role of the subordinated or the subaltern in theorizing the social is also a much discussed issue in anthropology, see e.g. Spivak 1988.

## 2.5 Conclusion: Invoking the Crowds

By invoking the 'public spectre' as the sudden rise of the unforeseen and eruptions of collective action, like Marres we suggest the need to revitalize the concepts of public and democracy once in the centre of John Dewey's moral and political philosophy (1927). But we argue instead that public formations in their concerns with issues occur from *communal* aspirations and concerns, however diffuse and unforeseeable, with felt social (in)justice, fear and anxieties and with rights of representation.

In this sense, we suggest to view the formation of publics as a process in time and in space, sometimes compressed and immediate, while at other times diluted, prolonged and diverted. The citizen action which we started out with is an instance of a sudden mass activity uniting individual fears and the terrible reminiscences of the past among inhabitants of Berlin. The public spectre becomes manifest in the eruptions of street action; in the second instance of the EPR the public spectre took form as a common concern with those made absent: the future end-users. To recapitulate, such empirical formations, the figure of the 'public spectre' as a prospect in social life evoke the image of a non-identical other or non-presence. When extraordinary events suddenly erupt and the unforeseen happens, the logic of negation threatens technological domination.

In this sense, we imagine the 'public spectre' in the same sense as the notion of the 'crowd' in the history of sociology was once captured (Borch 2012; Rudé 1964). In classic European thought from Gustave le Bon to Sigmund Freud, the crowd was essentially negative and a threat to public order. But with the evolution of sociological thought from Theodor Geiger to Herbert Blumer, the crowd assumed a positive image as it enabled isolated individuals to come together and express a collective voice regardless of social hierarchies. From this point of view, the crowd is a necessary, albeit not sufficient, step in the formation of the public in mobilizing sentiments and energizing collectivities. Marres suggests that publics are formed around neglected issues, we emphasize that issues unfold and are articulated in multiple and contested ways, on the grounds of shared concerns. Some concerns and issue articulations might remain at the margins of representation, evoking a public spectre.

John Dewey's legendary view of the 'public and its problems' wields much of the same sentiments as he fears 'organized politics' (with its orchestrated publics) to drain collective life in modern society of its energy. Crowds emerge spontaneously, and assume the figure of a 'public spectre' that resides in the unforeseen. When repeated as collective events, crowds stabilize and assume the figure of publics, i.e. physical inter-actions are mediatized around 'issues', with the eventual results that a more organized politics can emerge in terms of dialogues and compromises. But the public spectre – or the spectre of the crowd – will always haunt organized politics as its endemic 'other'; that the 'people' again will take action. The 'revolutionary crowd' was once an extreme spectre of the reign of chaos, while its modern successor as the many and diffusely erupting 'multitudes' help energizing the reified political present in promising people a public voice (Hardt and Negri 2000). So what we are proposing as 'the public spectre' haunting modern technological societies is the promise of 'radical democracy': that people can take action either as manifest or else as latent crowds. Such actions eventually threaten the prevailing order with chaos, and thus viewed as negative disruptions; nevertheless such disruptions also promise the rise of concerns, and thus issues, which in the absence of chaos would not otherwise arise. Whether realized or not, the spectre is always present as a (negative) potential that things can also go terribly wrong (as revealed by the Fukushima nuclear disaster). The promise inherent in technological change desperately needs a counter-figure; the public spectre (like the crowd in history) offers such a figure. The constructive purpose of such a (negative) figure resides in its possibilities to widen and enhance public engagement.

## References

Beck, U. 1992. *Risk society: Towards a new modernity*. London: Sage.
Blok, A. 2007. Experts on public trail: On democratizing expertise through a Danish consensus conference. *Public Understanding of Science* 16(2): 163–182.
Borch, Ch. 2012. *The politics of crowds: An alternative history of sociology*. Cambridge: Cambridge University Press.
Dewey, J. 1957 (1927). *The public and its problems*, 3rd ed. New York: Swallow Press.
European Commission (EC). 2001. *European governance – A white paper*. Brussels, 25.7.2001.
Gunnasdóttir, K., A. Mackenzie, and B. Wynne. 2011. *Social imaginaries and ethical issues in deliberative processes*. Introduction and summary of reports. Technolife Deliverable D-4.
Hardt, M., and A. Negri. 2000. *Empire*. Cambridge, MA: Harvard University Press.
Irwin, A. 2006. The politics of talk: Coming to terms with the "new" scientific governance. *Social Studies of Science* 36(2): 299–320.
Jasanoff, S. 2003. Technologies of humility: Citizen participation in governing science. *Minerva* 41: 223–244.
Jasanoff, S. 2005. *Designs on nature: Science and democracy in Europe and the United States*. Princeton: Princeton University Press.
Jasanoff, S. 2012. *Science and public reason*. New York: Routledge.
Jensen, C.B. 2005. Citizen projects and consensus building at the Danish Board of Technology: On experiments in democracy. *Acta Sociologica* 48(3): 221–35.
Latour, B. 2005. *Reassembling the social: An introduction to actor-network theory*. Oxford: Oxford University Press.
Leach, M., I. Scoones, and B. Wynne (eds.). 2005. *Science and citizens: Globalization and the challenge of engagement*. London: Zed Books.
Lezaun, J., and L. Soneryd. 2007. Consulting citizens: Technologies of elicitation and the mobility of publics. *Public Understanding of Science* 16: 279–297.
Lippmann, W. 1925/1993. *The phantom public*. New Brunswick: Transaction Publishers.
Marres, N. 2005a. A crucial proposition of the Lippmann-Dewey debate. In *No issue, no public: Democratic deficits after the displacement of politics*, 34–65, Doctoral dissertation, University of Amsterdam.
Marres, N. 2005b. Issues spark a public into being: A key but often forgotten point of the Lippmann-Dewey debate. In *Making things public*, ed. B. Latour and P. Weibel, 208–218. Karlsruhe/Cambridge, MA: MIT Press.

Marres, N. 2007. The issues deserve more credit: Pragmatist contributions to the study of public involvement in controversy. *Social Studies of Science* 37(5): 759–780.

Murphy, M. 2004. Immodest witnessing: The epistemology of vaginal self-examination in the US feminist self-help movement. *Feminist Studies* 30(1): 115–147.

Nascimento, Â., Â. Guimaraes Pereira, and A. Ghezzi. 2014. *From citizen science to do it yourself science: An annotated account of an on going movement*. European Commission, JRC.

Rudé, G. 1964/1981. *The crowd in history 1730–1848*. London: Lawrence and Wishart.

Scott, J. 1991. The evidence of experience. *Critical Inquiry* 17: 773–797.

Spivak, G.C. 1988. Can the subaltern speak? In *Marxism and the interpretation of culture*, ed. C. Nelson and L. Grossberg, 271–313. London: University of Illinois Press.

Welsh, I., and B. Wynne. 2013. Science, scientism and imaginaries of publics in the UK: Passive objects, incipient threats. *Science as Culture* 22(4): 540–566.

Wickson, F., A. Delgado, and K. Kjølberg. 2010. Who or what is the public? *Nature Nanotechnology* 5: 757–758.

Wynne, B. 2006. Public engagement as a means of restoring public trust in science – Hitting the notes, but missing the music? *Community Genetics* 9(3): 211–220.

# Part II
# Governing the Human Body and Mind

## On Modification and Enhancement Technologies

# Chapter 3
# Cochlear Implantation: Exploring Technoscience and Designs on Citizenship

**Kim Sune Jepsen**

**Abstract** Around the globe more than 350,000 citizens live with a sophisticated hearing aid device called Cochlear Implant. In many countries CI is an institutionalized medical procedure for restoring deafened with a technological sense of hearing but the social challenges for users is a largely unexplored issue. Addressing this social dimension, the article explores how Cochlear Implant works as a boundary object between social worlds and how it carries designs on citizenship. Designs on citizenship is not only an issue of unexpected forms of Deaf resistance. On a more substantial level, designs on citizenship concerns the way Cochlear Implant intersects with different social worlds. In specific, the articles explores the world of medical knowledge in conjunction with the social world of users. As new forms of living with technoscience is actively negotiated in the social world of usage, we may understand technoscience to carry designs on citizenship by opening up new places of belonging.

**Keywords** Technoscience • Cochlear implantation • Citizenship

---

K.S. Jepsen (✉)
Department of Sociology, Lund University, Lund, Sweden
e-mail: kim_sune.jepsen@soc.lu.se



A. Delgado (ed.), *Technoscience and Citizenship: Ethics and Governance in the Digital Society*, The International Library of Ethics, Law and Technology 17,
DOI 10.1007/978-3-319-32414-2_3

## 3.1 Introduction: Cochlear Implantation as a Social Technology

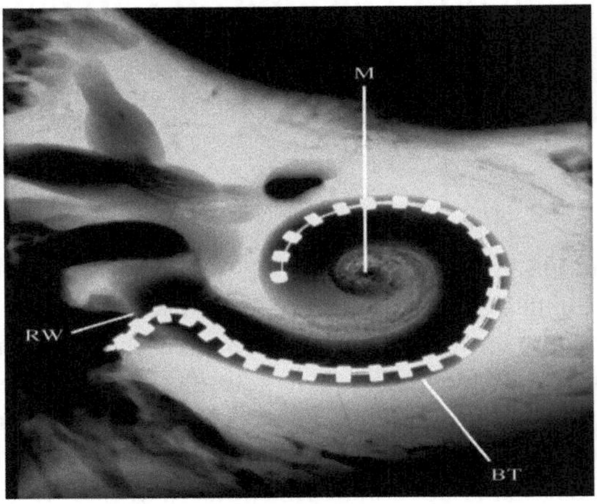


You are looking at the inner ear of a wired human being. The image depicts parts of a sophisticated medical technology called a Cochlear Implant. It is a device made to rehabilitate hearing disabled and deaf. A Cochlear Implant "*(bionic ear) restores useful hearing in severely to profoundly deaf people. It bypasses the inner ear and provides information to the hearing centers through direct stimulation of the hearing nerve*" (Clark 2003: xxxi)

Today many hearing impaired and deafened adults and children in Europe and the US have been implanted with a hearing aid device called Cochlear Implant (CI). The device was developed in the early 1970s at a stage of uncertain experimentation and it was subject to much scepticism among medical experts as to its feasibility of working and biological safety (Blume 2010). However, a few research communities did believe it was possible to restore a technological sense of hearing in deaf citizens. Notably in the US, UK and Australia research communities of medical experts, engineers, and designers began experimentations in the 1970s in order to bring 'Sounds *from* Silence' (Clark 2000).

In 1978 the first Australian implantation of an individual was a success. With the support of capital investments in the following years, the CI device established itself as a symbol of the powerful capabilities of emerging bionic technologies to modify the nature of disabilities. Today implantation is an ordinary institutionalized medical procedure in many countries with more than 350,000 implanted citizens around the globe. In the words of medical experts, CI is 'one of the most successful, life-transforming and expensive clinical interventions in modern medicine' (Tarabichi et al. 2008: 3).

Although CI is widely recognized as a life-transforming and commercial success, its institutionalization occasioned public controversies. In many countries deaf communities turned against the technology in the 1980s and 1990s. The life of deaf communities in hearing society has been well documented as a history of exclusion, which has enforced a strong cultural identity of a marginalized group (Harris 1995). Similar to how black struggles formed against oppression and turned stigma on its head by taking pride in being Black, Deaf people have organized around a shared culture. The statement in 1993 from the US National Association of the Deaf (NAD) regarding pediatric implantation exemplifies how issues of citizenship were invoked:"*Ortologists, speech and hearing scientists, manufacturers, parents, and members of the FDA staff were all consulted formally by the FDA in arriving at its decision. FDA's FAILURE to consult deaf spokespersons represents, if an oversight, GROSS IGNORANCE concerning growing up in deaf America, or, if willful, an offensive against fundamental American values of individual liberties, cultural diversity and consumer rights'* (NAD 1993). As STS scholar Stuart Blume has argued, such oppositions concerned the misrecognition of distinct cultural groups of citizens. CI became an object of resistance not least because it was commercialized under the signifier of a 'bionic ear' that promised liberation from a disability (Blume 1997).

At the core of such matters, Blume's case study of the nature of justice and politics entangled in the institutionalization of CI for children in the Netherlands is illuminating. Representing the Federation of Associations of Parents of Deaf Children (FODOK) on the subject of Cochlear Implantation in the 1990s, Blume was familiar with public resistance in countries like France, Britain and the US. Using his knowledge of how technoscientific interventions often carry uncertain social consequences and often involve disparate social worlds, Blume participated in organizing inclusive citizen forums. Such forums would make a political space possible in "which dialogue between the principal interest groups would be institutionalized." (Blume 2000: 160). Ethically, such forums would serve to constitute an 'ethical mean' through which differing social worlds of citizens could participate and be equally represented (Blume 2000: 145; 162).

Deaf people, medical experts, government representatives, social scientists, FODOK and interested others were brought together in the hope of reaching a policy crafted in consensus. Medical experts needed to apply for funding from the Health Insurance Counsel (HIC) to proceed and they wanted to avoid public confrontation. They were open to make a position statement taking other interested groups into consideration. A common statement was crafted. Medical experts emphasised the good and safe auditory improvements of implanted individuals. Deaf communities emphasised the recognition of their culture and that the use of sign language should be recognized, when informing future users of implantation. FODOK underlined that the unknown sociological and psychological consequences of implantation needed more and better assessment before institutionalizing Cochlear Implantation on a full scale. Contrariwise to this democratically informed recommendation that was send to HIC, the subsequent policy statement only recognized medical reasons. Indeed, the policy qualified this bias by arguing that it was a product of consensus. It did not recognize the different interests at stake. Subsequently, the citizen forums dissolved as the medical profession received its support to proceed on its own terms. As Blume underlined, the politics of CI strongly curtailed social justice because consensus became a resource that was used instrumentally.

This short story of CI exemplifies how the development of science and technology intersects with issues of social belonging and citizenship. CI concerned the technological rehabilitation of damaged regions in the inner ear but the technology became entangled in political disputes between different groups on a societal level. This points towards an important social dimension that has also been aptly formulated by the European Group on Ethics: *"The technological drive to promote cochlear implants raises ethical questions and concerns with how this drive impacts on the individual and on the deaf community…It does not pay sufficient attention to the psychological, linguistic and sociological issues. Above all it promotes a particular view of "normality"* (EGE 2005: 33).

The current article explores some of the above issues as it looks closer at how CI carries designs on citizenship. The theoretical proposition is that CI is not adequately grasped as a medical technology for disabled individuals. Rather, CI involves different social worlds and unresolved issues of what it means to live as a citizen with a radical technological innovation. I, thus, propose that we understand CI as a boundary object that intersects with different social worlds: *'…an object which lives in multiple social worlds and which has different identities in them'* (Star and Griesemer 1989: 409). Originally, this concept signified how different epistemic worlds of science are brought together in practices of translating and making an object work in distinct environments. More generally, boundary objects refer to junctures between different social worlds organized by collective practices, shared 'universes of meaning', and social interactions that demarcate the boundaries of a social world. This definition allows us to explore how a thing like a technology is made to matter in social worlds and how controversies between social worlds unfold in *an arena* of differently entangled actors, translations, representations and social interactions (Clarke and Star 2008).

Looking at CI as a boundary object, Blume's study exemplified how controversial social worlds and issues of citizenship were entangled. Understanding CI in such terms also underlines that boundary objects may not be restricted to collaborations between scientific worlds. As we shall see, the manner technology travels into social life and impacts on the life of users is an important lever for fashioning an understanding of other social worlds.

First, this article pursues a theoretical understanding of how CI carries *designs on citizenship*. With this latter notion I suggest that different social worlds are part of making CI matter and that certain ways of belonging to a social world are tied to technoscientific objects. The suggestion is, thus, that citizenship should not be restricted to obligations and legal, social and political rights of individuals within the nation state. Besides such crucial characteristics, new notions of citizenship are needed (Siim 2013) and I look at citizenship as a matter of belonging to social worlds entangled in technoscience. Second, the latter understanding is, especially, explored in the junctures between the social world of medical expertise and the social world of users. On this basis, the political and ethical nature of CI as a case of technoscience will be discussed.

## 3.2 The World of Technoscience: Designs on Citizenship

As technoscience increasingly interferes with our social existence, present us with technological possibilities for changing our life, alternative meanings of citizenships are emerging. A radical suggestion was, thus, made by cultural theorist Donna Haraway: '*We are all chimeras, theorized and fabricated hybrids of machine and organism; in short, we are cyborgs. The cyborg is our ontology; it gives us our politics.*" (Haraway 1991a: 150) Haraway suggested the provocative image of the cyborg to connote the problematic ways that the human is increasingly recast as an object of science, industry and state power. The trope of the cyborg was a rejection of the often widespread modernist belief in the teleological progress of science and technologies to simply identify and solve social problems. Contrariwise, the advent of technoscience was theorized to imply new ways of understanding, categorizing and objectifying what it means to be human. In turn, this understanding called for exploring the ways technoscience is situated in deeply cultural, political and social practices of engineering and transforming ways of living and belonging (Haraway 1991a). Taking into account that science carries cultural commitments and that citizens are differently affected, the methodological focus should be on uncovering 'situated knowledge': "…a collective subject position that promises a vision of the means of ongoing finite embodiment, of living within limits and contradictions, i.e. of views from somewhere." (Haraway 1991b: 196)

Haraway's proposal helps to situate citizens' life in the context of technoscientific change. In this respect, Science and Technology Studies (STS) have done much to uncover how natural science and forms of knowing are performatively ordering social life. As suggested by the framework of 'co-production', scientific expert knowledge materialized in technologies carry cultural assumptions and orders of representational politics in the form of social identities, discourses and institutional effects on social life (Jasanoff 2004). In conjunction, an important strand of STS has demonstrated how technological change is ordinarily framed by expert understandings and discourses that impact on the nature of civic life. To understand technoscientific change, we should not only look at how technical knowledge and skills are inscribed in a material object but also at how cultural assumptions constrain the ways in which 'citizens' and 'publics' can be concerned and adapt to technological innovation and change (Welsh and Wynne 2013; Wynne 2006). Following such analyses, we must look at the ways in which technologies are shaped, standardized and made to matter in social life by also opening up and producing boundaries on manners of belonging to social worlds. So, how might designs on citizenship look like?

An interesting proposition is that life science and technologies are changing the nature of disabilities and creating new forms of biological citizenship (Rose and Novas 2008). As new life sciences are changing disabilities, the subjectively experienced body is affected. On the one hand, new forms of citizenship are thus tied to new techniques and expert knowledge for curing bodily disabilities. On the other hand, new 'collective subject positions' may emerge in so far citizens identify with their technologically affected 'disabilities' in novel ways. Hence, technoscience may involve designs on citizenship in terms of how bodies are affected and new places of belonging are emerging (as also exemplified with the case of Cochlear Implantation).

In addition, anthropological inquiries into the globalization of the nation state have further uncovered new ways in which social identities, cultures and citizenship are emerging (Gupta and Sharma 2006; Li 2005; Sheppard 2002; Trouillet 2001). Such inquiries have looked at the ways in which transnational institutions and cross-national flows of culture primarily embodied in the globalization of capital and goods intersect with *places of belonging*. As aptly suggested by Trouillet (2001), citizenship may thus also be changing: "*…in the many practices through which citizens encounter not only government but also a myriad of other statelike institutions and processes that interpellate them as individuals and as members of various communities*" (Trouillet 2001: 133). This does not mean that citizenship is no longer tied to nation states and the particular identities, rights and obligations associated with national belonging. But as also suggested by the notion of biological citizenship, state-like effects in terms of new communal belonging to social worlds *may also be* emerging at novel sites of technoscience.

The above conception brackets ordinary state governed notions of citizenship in order to emphasize the social ways of belonging that are tied to technoscience and the changing nature of disabilities. State-like effects are tied the how forms of scientific expert knowledge shape a particular technology by also implanting social orders. Yet, the way citizens occupy 'a collective subject position' may, arguably, also depend on the active adoption of technological innovations. Whereas the notion of interpellation suggests an automatic subjugation to a certain social order (the althusserian proposition), the statelike effects in terms of social order and places of belonging should not lead us to neglect the active practices of adopting technological changes.

In respect to the above, CI is a global technological innovation that interferes deeply with bodily disabilities through 'the application' of a social order, identities, representations and places of belonging[1] (Image 3.1):

[1] The following images are from the published annual reports (editorial) and can be accessed on the homepage of Cochlear Limited: http://www.cochlear.com/wps/wcm/connect/intl/about/investor/annual-reports/annual-reports. From top left: Annual Report 2013, Annual Report 2012; Annual Report 2011; Annual Report 2010. From Bottom Left: Annual Report 2009; Annual Report 2008; Annual Report 2007; Annual Report 2006.

**Image 3.1** As the images exemplify, the meaning of the current globalization of CI in the version of Nucleus Freedom is represented by the good life (The reader is referred to the website and to annual report for more information on corporate globalization). But rather than taken these prospects for granted, the next chapter addresses the social worlds entangled in the designs on citizenship

## 3.3 A World of Medical Expert Knowledge

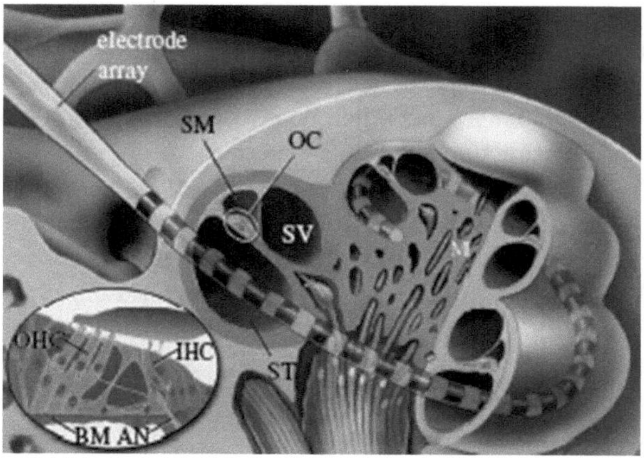


"The cochlea with 21⁄2–23⁄4 turns spiralling around the modiolus (M). The fluid-filled canals in the turns are: SV, scala vestibuli; SM, scala media; ST, scala tympani. Inset: the organ of Corti (OC) rests on the basilar membrane (BM). The OC has outer (OHC) and inner (IHC) hair cells connected to auditory nerve fibres (AN)" (Clark 2006: 792)

**Table I.** *Discrimination ability before and after implantation (from impossible = 0 to easy = 100) and frequency of conversation with others before and after implantation. Median value (2.5% and 97.5% percentiles)*

| Discrimination ability | Before | After | *p*-value |
|---|---|---|---|
| Discriminate a voice from background noise | 5 (0-70) | 85 (15-95) | <0.01 |
| Discriminate between a male and female voice | 5 (0-10) | 65 (5-95) | <0.01 |
| Understand normal speech without lip reading | 5 (0-40) | 50 (5-100) | <0.01 |
| How often did you talk to other people | 65 (5-95) | 95 (5-100) | <0.02 |

**Fig. 3.1** Faber and Grøntved 2000: 151

As explained, the social interventions of technoscience intersect with new ways of ordering and representing humanness and society. Hence, intersecting with the global trends of CI there is a growth in international medical assessments of life quality. But what is the character of expert knowledge as a social world? Answering this question, I will look closer at the medical assessments of adult CI-users in a Danish context. These medical assessments use common statistical approaches and refer to general findings in international medical journals.

When researching the international database pubmed, 78 articles concern the quality of life for adult CI-users and 4 are qualitatively grounded.[2] Danish clinical results are in consensus with international assessments that show a significantly good improvement in the quality of life. As exemplarily stated in an assessment of ten post-lingual deaf patients using their implant 16 h every day: 'The patients answered questionnaires dealing with communication abilities and quality of life, including 100 mm visual analogue scales, 1 year or more after receiving their CI' (Faber and Grøntved 2000: 151). In consensus with international trends, the statistical results are based on the categorization of specific social situations in order to evaluate the experience of CI users (Fig. 3.1).

As the figure shows, the results are significantly drawn towards 'easy' in the situations of evaluation. By the same token, it is clear that the quantitative evaluation is largely open to interpretation, e.g. it is unknown why some users value the post-implantation experience to 5 and precisely how many are below average (50) compared to users that evaluate the experience as high, e.g. 95. In effect, we do not know what kind of social challenges that might be involved in such variations. The methodological form of abstraction places the experience of users in a certain order of knowledge (Fig. 3.2).

The situation 'at home' also range from 'good' to 'easy' whereas 'amongst others' and being in a situation 'with background noise' reflect more variation and least satisfaction. Although the results indicate variation in CI-users experience, then an overall improvement is clearly present: 'CI dramatically changed the quality of life for all patients' (Faber and Grøntved 2000: 151). Such assessments exemplify a growing international consensus about the improvement of quality of life, which was formulated like this in a typical statistical assessment: "*The Glasgow Benefit*

[2] Only two of these are comprehensive inquiries into the experience of using CI: One uses an open-ended questionnaire, the other assessment makes use of grounded theory (Hallberg et al. 2004). Both assessments assert that CI has a good impact on the quality of life.

**Table IV.** *Satisfaction with CI use in different situations (from poor = 0 to excellent = 100). Median value (2.5% and 97.5% percentiles)*

| Situation | Satisfaction |
|---|---|
| At home | 95 (60–100) |
| At work[a] | 95 (55–100) |
| Amongst others | 80 (15–95) |
| Without background noise | 95 (50–100) |
| With background noise | 50 (0–70) |

[a]Three patients were unemployed and are not included in the calculation.

**Fig. 3.2** Faber and Grøntved 2000: 153

*Inventory revealed a positive effect in 93 % of patients. The use of a cochlear implant significantly enhanced discrimination ability, telephone use and self-confidence. A high degree of satisfaction was achieved in all situations except with background noise. Ninety-six percent of patients would recommend the operation to a friend. A dramatic improvement in quality of life following cochlear implantation is revealed by a great majority of patients. The results cannot only be explained by enhancements to auditory perception.*" (Lassaletta et al. 2006: 267).

As indicated above, the dominant form of representing CI-users experience is by means of statistical assessments.[3] While such medical assessments must be seen as a good start for closing the aforementioned 'knowledge gap' voiced by the European Group on Ethics (EGE 2005), they are largely based on statistical and positivist conceptions of how to measure life quality. In distinction, the notion of designs on citizenship leads us to understand how such technical assessments partake in the production and projection of a medical world and its orders of knowledge. Hence, the assessments represents a form of knowledge production that legitimize the technology by also framing a certain representative image of the user. Subsequently, how CI impacts on social life and what it means to live as a citizen with a technological sense of hearing is not dealt with in the medical world. In view of such knowledge deficits, the next chapter will address the emerging social world of users as it looks closer at the experience of five Danish adult CI users.[4]

[3]E.g. see also Krabbe, P.F., J.B. Hinderink, and P. Van den Broek. 2000; Grøntved, Aksel Møller et al (2006).

[4]The following paragraph is based on interviews with five post-lingual adult CI users used in a MA thesis (Jepsen 2008).

## 3.4 An Everyday Technological World

Medical assessments represent good outcomes of CI, but the deeper social challenges of using CI are not addressed. In this respect, an interesting phenomenological study has shown how *a technological sense of hearing* impacts deeply on social identity (Finlay and Molano-Fisher 2008). While Finlay and Fisher carried out a phenomenological exploration of how a single user experienced social challenges in the embodiment of a wired social world, such social issues of CI was also a common experience among the five interviewed post-lingual Danish users (in a study of the social integration of CI, Jepsen 2008).

The Danish users had lost hearing and for them deafness meant a rejection from a familiar social world:'*...but the worst change was when I couldn't understand speech, it was like being in glass bell and just to watch life without being able to participate.. I couldn't be like that anymore. And for many years I really felt like a bird without the tips of its wings: incapable of flying.'*[5] By the same token, CI was experienced as a liberating way out and a possibility to partake in social life again. However, the medical world tends to represent an image of how CI *provides* disabled individuals with a more or less ordinary sense of hearing. In distinction, users appear to experience persistent challenges in their everyday life: *'I can speak in the phone, but only with those I am familiar with', 'when people call I need to connect to speak', 'the birds still sounds strange, but it is alright because now I know, where the sound is coming from', 'I still avoid big gathering, because of the noise you can't really be in them', 'it is still best to face those you speak with', 'music still doesn't sound good', 'you get very tired in big gatherings'.*[6]

The above experiences reflect a continuous need to adjust oneself to a technological sense of hearing and its use in environments. As formulated by Ingrid: "*I think that hearing people can ignore sound. If it is in the background they don't notice. For me sound infiltrates loudly. Those sounds that are present will be analyzed and send to the computer.*"[7] Similar to the social dimension explored by Finlay and Fisher (2008), the users' experience testifies to how social identity is changed by acquiring a technological sense of hearing. A new sense of technological hearing involves a need to adjust oneself to social environments that are changing: *"I was at a concert with XXX who talks a lot in between the songs. Normally, I understand what he says, but I couldn't. And the other girls laughed and things. I thought what to do. Instead of laughing without any reason, I said to myself, I will not listen to XXX, but I will concentrate completely on this flute in the background. And I did it, and I got a completely different evening out if it then the others. But I succeeded and I learned that with the will I can. I have used it ever since."*[8]

---

[5] Jepsen 2008: 43.

[6] Jepsen 2008: 56.

[7] Jepsen 2008: 79.

[8] Jepsen 2008: 59.

The experience of a technologically changing social world is a common trait of all the users interviewed. In distinction to the medical world, the everyday social world of users involves ongoing challenges and distinct practices of actively shaping how a technological sense matters: "*…it is not always easy to understand others. I sometimes participate in guided tours in nature. Then I will just have to go up front to hear what is said; just say to people 'sorry, but I really don't hear that well' and then there is no problem. Say it – it will remove many obstacles on the road of the hearing impaired.*"[9]

## 3.5 Cochlear Implant: A Matter of Ethics and Voice

The above section reflects how a technological sense is adopted in conjunction with the challenges of engaging in a changing social world. On this account, it remains important to underline that the interviewees are adults that have lost their hearing. This group will properly have a different experience compared with those growing up with the implant. But on three accounts the experience of the interviewed group is useful for understanding a social dimension of CI.

First, the experience of users shows that CI does not restore the user to a normal sense of hearing. CI is a life-changing technology: it creates a new group of hearing impaired citizens that face the common challenge of being connected to a particular social world mediated by a technology. This should not make us neglect the good outcomes of CI. But we need to acknowledge that CI hangs together with new existential conditions situated in a social world between hearing and deaf.

Second, in some cases the institutionalization of CI has involved exclusion of Deaf concerns. However, it is also important to note that most deaf children have hearing parents that have a right to make the difficult choice of implantation. This right is grounded on an ethics of 'informed consent' but it also concerns a substantial choice between diverse life courses and social worlds. The ethical dimension of CI was also precisely formulated by a user. Lise said the following, when asked about Deaf resistance to the implantation of children:

> Lise: I think it is anxiety that makes them resist; the anxiety of losing them; that they will pass away and all that….
>
> Interviewer: But for me it is really difficult to comprehend this, because it is the anxiety of the unknown…
>
> Lise: yes, but we all have that. To a certain degree we all fear the unknown, because what we stand on is safe as we know it. It is when we have to walk across the edge, and it doesn't matter if it is a grim edge or less grim, we always have that moment of drawing a deep breath. It is natural…. But I think that one has to look beyond ones' own nose tip.[10]

Lise gave a reason for understanding deaf resistance but it may also point us towards a recognition of the existential uncertainties that normally accompany radi-

---

[9] Jepsen 2008: 58.

[10] Jepsen 2008: 96 (appendix 2).

cal technological changes of social life. In this respect, a democratic politics of new technologies must as a minimum take serious the inherent social uncertainties by recognizing how different groups are affected and allowed the right to voice concerns (see e.g. Wickson et al. 2010; Jasanoff 2003). As anthropologist Visvanathan (2005) has argued, the dispersed social consequences of technoscience calls for a democratic politics that does *cognitive justice* to forms of everyday knowledge (Visvanathan 2005: 92).

Third, the advertised representations and the manner medical discourses project a certain image of users tend to abstract from the ways in which CI is made to work in everyday life. Listening to users, we have gained an understanding of how technological hearing connects with the adjustment to a social world. As Lise said, when talking about the need to have a firm 'will':

> Lise: You can get a push towards this will. But you can't get it for free. It good to get support from educational teacher and friends, if there are problems or you are stuck
>
> Interviewer: So what kind of problems could they be?
>
> Lise: It doesn't necessarily mean to be stuck in respect to education, but it can also mean to be stuck with others in relation to oneself in relation to CI[11]

## 3.6 An Ethical World: Towards an Existential Politics

An ethical issue concerns the misrecognition and exclusion of Deaf concerns. Deaf resistance was grounded in a culture that turned a disability into a symbol of a particular social world and ways of belonging. This act placed value on peoples' right to have a specific minority culture of belonging. In distinction, a medical discourse has determined deafness as a disability that can be corrected by implants, thus, placing value on individual liberty and rights to freely choose.

Australian philosopher Neil Levy (2002) has provided an interesting clarification to solve the above ethical dilemma (the controversial meanings and values of Cochlear Implant). Levy identifies and evaluates three arguments in the discourse of Deaf resistance to CI: (1) deafness is not a disability. It originates in a society of discrimination; (2) treating deafness with CI sends the message that deaf citizens are of lesser worth; (3) deafness provides members with access to a rich culture and even ethnicity that should be acknowledged as intrinsically valuable (for members). This latter fact yields a moral obligation to protect deaf culture against eradication by CI.[12]

Levy contents the first argument by arguing that the implications of deafness are mainly of a social origin. Deafness depends on how institutions are structured (inequality in work life, education etc.) and we have no reason not to change such institutions in the case of discrimination. However, in some important situations, as in traffic, deafness contains some unavoidable disabilities that we have a compelling

[11] Jepsen 2008: 59.

[12] Lane 2005.

reason not to accommodate. Hence, in principle deafness can be subject to medical intervention (ibid. 141).

The second argument largely depends on whether or not CI interferes with a cultural identity or with a disability that is absolutely distinct from culture. This ultimately depends on the third argument which is satisfied when deafness constitutes a particular culture (Levy 2002: 144). In consequence, Levy argues that the ethical dilemma (should society provide Cochlear Implant or not) must be resolved by attributing different moral weight to the disadvantages suffered by deaf (argument 1) against (argument 2 and 3) the 'intrinsic value of deaf culture'. So how might we attribute moral value?

Levy's interesting solution to this question is found by invoking Nora Groces' (1988) ethnographical and historical research of deafness many decades ago on a small Island outside the American east coast, Martha's Vineyard. Through the recollection of old inhabitant's experience and documental research, Groce shows that in regions of high numbers of deaf people 'everyone spoke sign language': both hearing and deaf. When no communication barriers existed, then cultural life was not divided by handicap (Groce 1988: 4): "*The Martha's Vineyard experience suggests strongly that the concept of a handicap is an arbitrary social category.*" (Groce 1988: 108) Deafness on Martha's Vineyard was mutually adjusted by citizens. The islanders gained *a common experience* of living with deafness through a shared (sign) language in a particular social world.

Drawing on the example of Martha's Vineyard, Levy's argument is now straight forward. Avoiding social disadvantages of deafness comes with the cost of a significant price: "*…the death of deaf culture….since everyone spoke sign language they were fully integrated into the community – which is to say that there was no separate Deaf culture.*" (Levy 2002: 151). In the context of CI this, arguably, corresponds to giving moral weight to argument 1: '*The only practical way to reduce these costs* [disadvantages] *is through the assimilation of the Deaf, whether through the generalization of Sign, or the use of implants. Either the Deaf must continue to bear the costs of their disability or they must disappear*' (Levy 2002: 151 [my insertion])

While Levy's philosophical reflection serves to solve the ethical dilemma of CI, its resolution can be contested. Martha's Vineyard did not lead to the death of Deaf culture, because deaf citizens were not assimilated but integrated. This crucial political dimension is missed by Levy when he asserts that: "*deaf Vineyarders performed as well as anyone….The exception was education, on which the deaf tended to outperform their hearing neighbours.*" (Levy 2002: 141). Groces' study did not call for a politics of assimilation, but it was meant to show how deafness is not only biological but also an arbitrary category of societal organization.

The example of Martha's Vineyard suggests that as an arbitrary social category deafness can be negotiated and changed for the better. Related to this recognition the implementation of CI may be grounded on an ethics that makes differences a cause for more and better co-participation (this was also Blumes' suggestion of finding an 'ethical mean'). In this sense, an ethical principle of technoscientific innovations may concern a need for practical collaboration of diverse social worlds.

An informative example of the above is the study of the Association of French Muscular Dystrophy Patients (Rebeharisoa and Callon 2004). In this study, the innovation of specific tools among the patient groups allowed for the negotiation between personal everyday experience and formalized expert knowledge (Rebeharisoa and Callon 2004: 159). It was the intersections between otherwise distinct social world that made new democratic politics and useful integration possible. With specific technical tools patients and relatives were enabled with possibilities to negotiate with laboratory experts and secure collaborative engagements in making the nature of disease and disabilities matter.

In a similar vein, this paper proposes that the experience of CI-users reflects a need to recognize the social dimension of technoscience and how it carries designs on citizenship. CI is currently credited with revolutionary capabilities to change social life, but the dominant order of expert knowledge largely abstracts from the ways in which citizens-as-users are living with a technological sense of hearing in a social world.

## 3.7 Conclusion: Technoscience, Social Life and Citizenship

The case of Cochlear Implantation is an example of how technoscientific innovations are deeply entangled in new ways of belonging to a social world. Whereas citizenship is often theorized within the framework of nation states, this article suggests a supplementary understanding. *Designs on citizenship* are intimately tied to how social belonging is affected and social worlds are shaped by new technologies. The case of Cochlear Implantation reflects how technoscientific innovations unexpectedly provoke public controversies and how technology is made to work on social grounds that are different from the dominant social world of expert knowledge. This is not to contest the value of the latter. But we need to acknowledge that the changing social world of users remain important for understanding the social implications of technoscience on how citizenship may be changing.

In particular, designs on citizenship have been suggested to concern the changing nature of disabilities. In this respect, this article has tried to explicate the arena of intersecting social worlds of expert representations, negotiations and usage entangled in how CI carries designs on citizenship. We have gained an understanding of how hearing impairment is not so much eradicated as substantially changed by the use of CI. Arguably, CI carries the emergence of a social world of living with technology between hearing and deaf worlds. Yet, future studies are needed to address how new needs and demands of citizens may be emerging in relation to technologically changing disabilities and the social worlds of usage.

Finally, it has been suggested that it is by bridging social worlds, and associated social identities, that new democratic integrations of technoscience can be achieved. As indicated, it is not least the common embodied experience of users that make it possible to uncover a "collective subject position" of technologically affected citizens. Citizens become imbricated in emerging social worlds and identities of a peculiar character as technoscience and its associated worlds of expertise make new places of belonging possible.

## References

Blume, S. 1997. The rhetoric and counter-rhetoric of a "bionic" technology. *Science, Technology & Human Values* 22(1): 31–56.
Blume, S. 2000. Land of hope and glory: Exploring cochlear implantation in the Netherlands. *Science, Technology & Human Values* 25(2): 139–166.
Blume, S. 2010. *The artificial ear: Cochlear implants and the culture of deafness*. New Brunswick: Rutgers University Press, January 2010.
Clark, G. 2000. *Sounds from silence – The bionic Ear story*. Australia: Allan & Unwin.
Clark, G. 2003. *Cochlear implants: Fundamentals & applications*. New York: Springer.
Clark, G. 2006. The multiple-channel cochlear implant: The interface between sound and the central nervous system for hearing, speech, and language in deaf people—A personal perspective. *Philosophical Transactions of the Royal Society of London. Series B, Biological Sciences* 361: 791–810.
Clarke, A., and S.L. Star. 2008. The social worlds framework: A theory/methods package. In *The handbook of science and technology studies*, ed. E.J. Hackett, O. Amsterdamska, M. Lynch, and J. Wajcman, 113–137. Cambridge, MA: Massachusetts Institute of Technology.
Faber, C., and A. Grøntved. 2000. Cochlear implantation and change in quality of life. *Acta Oto-Laryngolica. Supplementum* 543: 151–153.
Finlay, L. and P. Molano-Fisher 2008. Transforming' self and world: A phenomenological study of a changing lifeworld following a cochlear implant. *Medicine Health Care and Philosophy* 11(3): 255–267.
Groce, N. 1988. *Everyone here spoke sign language: Hereditary deafness on Martha's Vineyard*. Harvard University Press.
Grøntved, A. et al. 2006. Cochlear implantation in deaf adults: Effect on quality of life. Co-authored with Jens Højberg Wanscher, and Christian Emil Faber. Ugeskrift for læger 168/33, Videnskab og Praksis.
Gupta, A., and A. Sharma. 2006. Globalization and postcolonial states. *Current Anthropology* 47(2): 277–307.
Hallberg, L. et al. 2004. Living with cochlear implant: Experience of 17 adult users in Sweden. Co-authored with Anders Ringdahl. *International Journal of Audiology* 43: 115–121.
Haraway, D. 1991a. A cyborg manifesto: Science, technology, and socialist-feminism in the late twentieth century. In *Simians, cyborg and women*, Free Association Books LTD 1997.
Haraway, D. 1991b. Situated knowledges: The science question in feminism and privilege of partial perspective. In *Simians, cyborg and, women*, Free Association Books LTD 1997.
Harris, J. 1995. *The cultural meaning of deafness*. England: Avebury Ashgate Publishing Limited.
Jasanoff, S. 2003. Technologies of humility: Citizen participation in governing science. *Minerva* 41: 223–44.
Jasanoff, S. 2004. The idiom of co-production. In *States of knowledges – The co-production of science and social order*, ed. S. Jasanoff. Routledge.
Jepsen, K. 2008. *Doing sociological research and travelling boundaries of cochlear implantation – To learn from experienced people using CI how a sense is actively shaped*. Unpublished MA-dissertation, Lancaster University.
Krabbe, P.F., J.B. Hinderink, and P. Van den Broek. 2000. The effect of cochlear implant use in postlingually deaf adults. *International Journal of Technology Assessment in Health Care* 16(3): 864–873.
Lane, Harlan. 2005. Ethnicity, ethics, and the deaf world. *Journal of Deaf Studies and Deaf Education* 10(3): 291–310.

Lassaletta, Luis et al. 2006. Quality of life in postlingually deaf patients following cochlear implantation. Co-authored with Alejandro Castro, Marta Bastarrica, Maria Jose´ de Sarria´, and Javier Gavila´n. *European Archives Oto-Rhino-Laryngology* 263: 267–270.

Levy, N. 2002. Reconsidering cochlear implant: The lesson of Martha's Vineyard. *Bioethics* 16(22).

Li, T. 2005. Beyond the state and failed schemes. *American Anthropologist* 107(3): 383–394.

National Association of the Deaf. 1993. Cochlear implants in children. Position Statement from 1993.

Rebeharisoa, V., and M. Callon. 2004. Patients and scientists in French muscular dystrophy research. In *States of knowledges – The co-production of science and social order*, ed. S. Jasanoff, Routledge.

Rose, N., and C. Novas. 2008. Biological citizenship. In *Global assemblages: Technology, politics, and ethics as anthropological problems*, ed. A. Ong and S.J. Collier, 439–469. Oxford: Blackwell Publishing Ltd.

Sheppard, E. 2002. The places and times of globalization: Place, scale, networks and positionality. *Economic Geography* 78(3): 307–330.

Siims, Birthe. 2013. Citizenship. In *The Oxford handbook of gender and politics*, ed. Georgina Waylen, Karen Celis, Johanna Kantola, and S. Laurel Weldon, 1–18. Oxford University Press.

Star, S., and J.R. Griesemer. 1989. Institutional ecology, 'translations' and boundary objects: Amateurs and professionals in Berkeley's Museum of Vertebrate Zoology, 1907–39. *Social Studies of Science* 19: 387–420.

Tarabichi, M.B. et al. 2008. Deafness in the developing world: The place of cochlear implantation. Co-authored with C Todd, Z Khan, X Ýang, and B Shehzad. *The Journal of Laryngology and Otology*.

The European Group on Ethics, EGE. 2005. *The ethical aspects of ICT implants in the human body*. Luxenburg: Office for Official Publications of the European Communities.

Trouillet, M-R. 2001. The anthropology of the state in the age of globalization. *Current Anthropology* 42(1): 125-138.

Visvanathan, S. 2005. Knowledge, justice and democracy. In *Science and citizens: Globalization and the challenge of engagement*, ed. M. Leach, I. Scoones, B. Wynne. Zed Books Ltd.

Welsh, I., and B. Wynne. 2013. Science, scientism and imaginaries of publics in the UK: Passive objects, incipient threats. *Science as Culture* 22(4): 540–566.

Wickson, F., A. Delgado, and K. Kjølberg. 2010. Who or what is the public? *Nature Nanotechnology* 5: 757–758.

Wynne, Brian. 2006. Public engagement as a means of restoring public trust in science hitting the notes, but missing the music? *Community Genetics* 9(3): 211–220.

# Chapter 4
# The Modification of the Human Body: Controversies

**Søren Holm**

**Abstract** This chapter provides an overview of the academic debate about whether or not we should limit peoples' ability to pursue bodily modification and relates this to important results from the Technolife project. It identifies four major controversies in the debate: (1) Is freedom the Master Value of society? (2) Is there a valid distinction between treatment and enhancement? (3) Is enhancing only the rich unjust? (4) Are there any limits at all?

For each of these controversies the major arguments on each side are outlined and evaluated. It is concluded that society does have a role to play in determining limits to bodily modification, but that that role is only legitimate if there is public engagement.

**Keywords** Body modification • Enhancement • Justice • Liberalism • Nature • Natural law • Therapy • Transhumanism

## 4.1 Introduction

As a human being I don't have a body, I am embodied. We can't meaningfully distinguish between me and my body as Cartesian dualists want to do. This means that practices aimed at modifying the body often become controversial and that they often raise complicated ethical and policy questions, because modifying my body is inseparable from modifying the person who I am. Certain body modifications will not only change my body, they will also over time change me.

Among the questions are: Should there be limits to what kind of bodily modifications we allow? Does this depend on a distinction between modifications that are treatments and modifications that are enhancements? Does it depend on the methods for modification or the context in which they happen? Who should decide on bodily modifications in children?

S. Holm (✉)
Centre for Social Ethics and Policy, University of Manchester, Manchester M13 9PL, UK
e-mail: Soren.holm@manchester.ac.uk

 
A. Delgado (ed.), *Technoscience and Citizenship: Ethics and Governance in the Digital Society*, The International Library of Ethics, Law and Technology 17,
DOI 10.1007/978-3-319-32414-2_4

Answering these questions is further complicated because bodily modification takes place in many, partly disconnected areas of social life including medicine, sport, fitness, tattoo and piercing establishments and beauty parlours, to mention just a few. Each of these has its own set of practices and 'rules' for bodily modification. Each has its own community of practitioners, users and participants in discourse. And each has its own attendant academic literature spanning applied ethics, the social sciences, law and the humanities more broadly. There are for instance large literatures on both cosmetic surgery and doping in sport, but these literatures are largely unconnected even though both practices involve bodily modification. Each of the various areas of body modification is also exploited in science fiction and other literary or cinematographic works. This chapter will outline some of the controversies that cut across the various areas of body modification with the aim of identifying and discussing the factors that create controversy.

Getting answers to these questions is important. For instance, in sport we need to know why doping is wrong when it is wrong so that we can design appropriate anti-doping programs. And we need to know whether Oscar Pistorius, the double amputee 'Blade Runner' should have been allowed to compete in the Olympics: was his bionic prosthesis an enhancement that gave him an advantage over his competitors? What is the fair thing to do in this case? For our society we need to know whether we should pursue the dream of making our bodies impervious to disease, or whether we should perhaps instead work to leave our bodies behind and upload ourselves to computers. And in relation to our children and the children of others we need to decide what body modifications parents can legitimately have done (Ramsey 1970). Is it, for instance acceptable to have your male child circumcised, or to have your daughter's ears pierced?

## 4.2 The Body and Its Limits

How should we conceive of the body in ethical argument? And, does the present state of the human body and its current limit have any relevance as a premise in such arguments? Is there, for instance any normative force in the human life span presently being limited to approximately 120 years? The controversy concerning whether the body is normative or not is closely parallel to the even more general controversy in ethics around whether nature or the natural has any normative status.

The idea that nature is normative[1] is often linked to a more general adherence to religious or secularised versions of natural law theory (Finnis 1980; Farley 2001; Oderberg and Chappell 2004). Within this tradition nature or the natural is seen as normative because nature, properly interrogated reveals an underlying and meaningful order. What we see in nature is not just a random result of meaningless biological processes, but a meaningful whole that can guide us in our thinking about

[1] Normative: implying, creating, or prescribing a norm.

how we ought to live and act. A similar view can also be given a non-religious expression. What is good for human beings, what their interests are and what it is wrong to do to them is at least partly decided by their biological properties. Causing pain is, all other things being equal ethically reprehensible because human beings can feel pain and because pain has a strongly negative phenomenological quality. If we lived in a world where most agents did not feel pain, or where pain was only occasionally linked to strong negative feelings, then our ethical judgements might be different. Or if our way of having children involved us abandoning our eggs to their fate, as in many fish and reptiles then our considerations concerning reproductive rights and freedoms would have to be different. So biology does matter in some way. What is important is to decide how much and in what ways.

It is also obvious that much of the argument for enhancement (implicitly) assumes that features that are characteristic of current human bodies and beings will also be characteristic of future enhanced agents (Holm 2006, 2007). It is, for instance assumed in almost all of the pro-enhancement argument that any future agents will have the same kind of interest in freedom as human beings have.

On the other side of this debate it is denied, following the Scottish philosopher David Hume that nature, whether human or non-human nature can have any normative import. This point of view is for instance articulated forcefully by John Harris and by Tuija Takala among many others (Harris 2007; Takala 2004). The main arguments on this side of the debate is first that nature is just a brute fact and that the so-called fact/value distinction or the is/ought gap entails that no ethical judgement can follow directly from such a brute fact. Second that even if we took nature as normative we ought to realise that much of what is natural is not good and that most of medical activity is directed at preventing the natural from taking place. It is, it is claimed, natural for human beings to fall ill, suffer injury and die prematurely; and unnatural to use the products of highly advanced chemistry or the intricate procedures of modern surgery to combat these natural events.

This philosophical debate is unlikely to be settled any time soon.

Primarily outside of the academic literature there is a lively internet discussion of body modification on web-sites such as www.bmezine.com, www.bodymod.org and www.bodymods.org. What is discussed in the forums on these sites are, unsurprisingly not issues about the normative status of the human body. The body modifications that are described, pictured and discussed on these sites and their associated blogs and forums are mainly sought as means of personal expression and include tattoos, piercings, surgical modification and implants, scarification etc. Discussions concerning limits of body modification mainly occur in response to particular instances of body modification that are displayed and often focus on issues of aesthetics, long term durability and safety. There are, however some instances where the question of limits is explored more directly. In the academic ethics literature discussion of this kind of body modification which primarily takes place outside of a medical context is almost absent, although the prominent Swedish Transhumanist Anders Sandberg has argued that a right of 'morphological freedom', i.e. a right to modify your body in any way you choose is a basic human right (Sandberg 2001).

The deliberative exercise on body modification that was part of the Technolife project was one of the most active and animated discussions within the project, with a wide range of participants from across the world (Strand et al. 2011). Many of the issues that are discussed in the academic literature also surfaced in that discussion, but some other issues were raised as well, and these will be mentioned and discussed when relevant.

## 4.3 Reasons, Justifications and Controversies

What are the reasons given for pursuing modification and what are the justifications invoked for either allowing or limiting specific modifications?

One prominent reason is that the modifications that are sought are enhancements, i.e. that they in some way constitute an improvement of the body and the person. The person who has a tattoo probably believes that he looks better with the tattoo as does the person who goes to a cosmetic surgeon to have breast enlargements or a nose job. And in sports the reason for having your body modified by doping is to achieve some kind of enhanced performance.

But how should we understand the concept of enhancement? In a recent paper Holm and McNamee have argued that there are four principal ways in which this notion can be understood (Holm and McNamee 2011). There are two distinctions that generate the four different options. The first distinction is concerned with who is to judge whether a certain change is an improvement. Thus we can ask of any purported enhancement whether it is determined objectively (or inter-subjectively at least) or whether it is exclusively a matter for personal judgment. The second distinction is concerned with what the proper baseline is for judging or measuring improvements. Is it, for instance a species-based norm, or is it the function or appearance of one specific individual, or is there no norm at all but only a question of pure personal preference. This second distinction also plays a role in the debate concerning whether we can meaningfully distinguish between therapy and enhancement, because one way of defining therapy is as bringing someone back to normal functioning.

These two intersecting distinctions give rise to the four different ways of (mis) understanding improvement and *a fortiori* enhancement. The most restrictive of these is the objective, norm-based account which usually underlies arguments making a distinction between therapy in the sense of bringing a person (or restoring them) to a normal level of functioning and enhancement in the sense of improving function above the normal (Daniels 2009). The most expansive account is the subjective, status quo account which essentially entails that any physical change a person deems to be an improvement of his function is an improvement (Harris 2009).

The four options can be illustrated in the following way (Table 4.1):

**Table 4.1** Evaluation of bodily changes

| Improvement | Norm-based evaluation | Status quo based evaluation |
|---|---|---|
| Objective (or inter-subjective) evaluation | *Most restrictive account* | An improvement is anything that improves the individual in relation to his or her present state, as judged objectively by others |
| | An improvement is anything that improves the individual in relation to an external norm, as judged objectively by others | E.g. The sports "doping" account |
| | Treatment/enhancement distinction | |
| Subjective evaluation | *Rarely held* | *Least restrictive account* |
| | | An improvement is anything that improves the individual in relation to his or her present state, as judged subjectively by him- or herself |
| | | E.g. Often implied in trans-humanist writing |

Each of the four basic accounts can be further refined. As noted above improvements can be specific or general in relation to function, but it is also possible to conceive of improvements of the body which do not strictly improve any particular function yet still improves the welfare of the person. An improvement that makes you more beautiful is not strictly a functional improvement, but it may still increase your welfare (or it may not).

Within the subjective accounts further distinctions can be made in relation to whether the subjective judgment is personal or social and whether it can be fallible or not. This distinction may be important in relation to whether body modifications can legitimately be restricted. An argument for a human right of morphological freedom is much easier if persons are only rarely fallible in relation to what is in their own interest. Many body modifications are long lasting or permanent or can only be reversed at considerable cost or with considerable difficulty. If people are often wrong about whether such a modification is an improvement we may pause before allowing complete freedom in the body modification area.

In the Technolife discursive forum on body modification a frequent trope was to analogise body modification and other enhancement technologies to developments in ICT, either to show the great potential for development, or to direct attention to the potential problem of obsolescence. ICT has developed very rapidly, but part of this development has involved older devices and software eventually becoming obsolete, and participants predicted obsolescence of bodily modification and worried about the implications when things that are inside the body and not as easily exchangeable as your PC or your mobile phone become obsolete.

### 4.3.1 Controversy 1 – Freedom as the Master Value of Society?

This points to a general controversy which permeates the academic literature on body modification, what should be the master value or philosophy of modern society. Is modern society liberal or libertarian (or if not fully yet, should it be)? Is it, for instance acceptable for society to legally prohibit people from having tattoos on their hands and faces, as in Denmark? Or is it acceptable to legally limit some kinds of plastic surgery? If it is accepted that liberalism is, or should be the foundational value of modern society it becomes unproblematic to invoke John Stuart Mill's so-called 'harm principle' in aid of enhancements that do not harm others:

> The object of this Essay is to assert one very simple principle, as entitled to govern absolutely the dealings of society with the individual in the way of compulsion and control, whether the means used be physical force in the form of legal penalties, or the moral coercion of public opinion. That principle is *that the sole end for which mankind are warranted, individually or collectively, in interfering with the liberty of action of any of their number, is self-protection. That the only purpose for which power can be rightfully exercised over any member of a civilized community, against his will, is to prevent harm to others. His own good, either physical or moral, is not a sufficient warrant.* He cannot rightfully be compelled to do or forbear because it will be better for him to do so, because it will make him happier, because, in the opinions of others, to do so would be wise, or even right. These are good reasons for remonstrating with him, or reasoning with him, or persuading him, or entreating him, but not for compelling him, or visiting him with any evil, in case he do otherwise. To justify that, the conduct from which it is desired to deter him must be calculated to produce evil to some one else. The only part of the conduct of any one, for which he is amenable to society, is that which concerns others. In the part which merely concerns himself, his independence is, of right, absolute. Over himself, over his own body and mind, the individual is sovereign. (Mill 1978, my emphasis, p. 9)

But even if the harm principle is accepted as blocking prohibition of acts that do not harm others, Mill and the principle still allows for remonstration, reasoning, persuasion and entreating as legitimate means of restricting the actions of those who harm themselves. Let us consider what might constitute legitimate remonstration. There are two different kinds of remonstration depending on which flaw in the person or his thought processes we are pointing out. Let us call it prudential remonstration when we are saying to someone that he is foolish or reckless (in a non-moral sense of recklessness) and moral remonstration when we are saying that his decision is morally wrong and perhaps even that he would be a morally bad person if he acts on his morally wrong decision.

Both kinds of remonstration should be unproblematic to the liberal (even if Mill had not allowed remonstration) since they involve no compulsion, coercion or use of force. However, moral remonstration has fallen distinctly out of favour. It is no longer considered good form to tell people that what they are doing is morally wrong, if their action is purely, or even mainly self-regarding.

A whole range of actions that fall within the scope of a principle of morphological freedom seem to be exactly the kind of actions that are open to both prudential and moral remonstration. They are foolish, morally problematic or both; and both problems often can be traced to the same root course, that a person can only see the action in question and the bodily modification it will lead to as desirable because of a neglect or excessive discounting of long term effects.

For this class of modifications pointing out to a person that 'you will regret doing this' and giving reasons why this will be the case is just helping him or her to make rationally better decisions (it may be conceived as a sub-type of prudential remonstration). If, in addition some of the regret will be moral regret, for instance because future bad consequences affected others than the person in question, telling the person to think again and more clearly should also lead to morally better decisions.

Remonstration or persuasion may also be based on a notion of authenticity. In the present context the notion that my personality has a relatively stable core and that some of my autonomous decisions are more congruent with that stable core than others are. Defining exactly what that stable core is, and for how long it is stable is not easy and I will not attempt it here. Because even under conditions of liquid (post-)modernity it makes sense to make the remark in certain circumstances that 'acting in this way seems so unlike you'. The person in question may then respond that she is currently redefining who she is, or that there are parts of her person that I don't know and thereby close of further discussion. But she might also say 'why do you say that?' and engage in discussion. If she does that there does, again seem to be no problem in explaining to her that her actions seem to be against her prudential or moral values and that she should think again.

All of this seems to indicate that taking actions to ensure that a person is aware of what we could call the 'rational regret potential' of his or her actions is a morally acceptable or even good thing to do.

Even if we accept that both moral and prudential remonstration are acceptable under certain circumstances we could claim that only remonstration performed by individuals is acceptable, i.e. that collective agents like the state should not be allowed to remonstrate with those whose actions have considerable regret potential. Is this a sustainable position?

There does not seem to be any reason in principle why the state should not generally be able to remonstrate with its citizens. It does, for instance seem perfectly acceptable to for the state to put the message 'Stop smoking, only stupid people or those with crazy temporal discount factors smoke' on all cigarette packs (whether it would do the job is another matter). But perhaps the worry is not about this very general form of remonstration, but about more specific and individualised remonstration, or about remonstration requiring more direct compulsion of the individual. Here it is important to distinguish three possible strands of argument: (1) that the remonstration will be problematically biased, (2) that the mere existence of individualised remonstration will have a chilling effect on those seeking to have bodily modifications, and (3) that the mere existence of individualised remonstration will introduce a problematic time delay.

The third of these strands of argument is clearly relevant in cases like access to abortion where it has been discussed extensively in the USA. But a time delay in the case of many bodily modifications is not nearly as serious (whether I have my Botox today or next week will not matter much in most cases).

The first two strands are connected since (negatively) biased remonstration is likely to have a more chilling effect than neutral remonstration. So let us try to disentangle them by looking at two questions: (1) can a state legitimately engage in neutral remonstration, and (2) if a state can engage in neutral remonstration can it still do so if neutral remonstration is shown to have a chilling effect with respect to a certain class of acts.

Let us for instance imagine that we introduced compulsory counselling for anyone planning to have their face tattooed, and that this counselling was neutral and purely exploratory in the sense that it explored whether the person in question had considered the positive and the negative effects of having such a tattoo, now and in the future. This service would be provided for free and at the end of the counselling the person would be free to have his or her tattoo if they still wanted it. It is difficult to see that this would have a chilling effect or that the state would overstep any moral boundaries.

### *4.3.2 Controversy 2 – Is There a Difference Between Treatment and Enhancement?*

A further area of controversy is whether there is a morally relevant distinction between medical treatment on the one hand and bodily enhancement using the same technologies or skills on the other hand. Is there, for instance a difference between the plastic surgeon correcting congenital defects or burn scars and the same surgeon enlarging breasts that are of normal size? (see for instance the papers in Beaufort et al. 1996). Or is it really the proper work of dentists to make our teeth straighter, whiter and more beautiful?

This debate can be found in the literatures on cosmetic surgery, doping in sports, genetic enhancement, cognitive enhancement etc. (Parens 1998; Murray 2006).

Those who claim that the distinction is valid and useful usually do so by arguing that treatment for *bona fide* disease states has a moral claim on us that is stronger than the claim generated by enhancement that merely adds to a healthy individual's welfare. The *locus classicus* for this argument is the works of Norman Daniels. Daniels argues that the distinction is valid and useful because it reflects the basic principle of social justice that society should offer 'equality of opportunity' to all its members (e.g. Daniels 1985, 2009). Illness and disease prevent persons from achieving equality of opportunity and we therefore have a strong obligation (as a society) to prevent, treat and alleviate these conditions. Not being enhanced does not affect equality of opportunity in the same way (and there may even be an equality of opportunity based argument against enhancement if it is unequally distributed).

Those who claim that the distinction is invalid or without normative importance argue (1) that the standard accounts of illness and disease, for instance in terms of deviation from species typical functioning are highly problematic and (2) that it is difficult to see why certain kinds of improvement in welfare (those we call treatment) are more important than other kinds (those we call enhancement) (Harris 2009; Buchanan 2011). They further argue that even if the distinction is valid it provides no reason to ban enhancements; it only gives a reason to give them lower priority when we are distributing societal resources.

### 4.3.3 Controversy 3 – Is Enhancing Only the Rich Unjust?

The third major area of controversy in the literature centres on the question of whether there are problematic justice implications of body modification and enhancement. It is generally accepted in the literature that some enhancement technologies are likely to be used first by resource rich nations and/or resource rich persons and that when such technologies are introduced there will be a phase where they and their enhancing results are therefore unequally distributed. In the context of enhancement via genetic engineering Silver for instance predicts a future where some people are 'gene rich' and others are 'gene poor' (Silver 1998). Some might have bodies that are resistant to a wide range of common infectious diseases, while some will still be susceptible, or some might have hearts that repair themselves much more effectively after heart attacks.

The questions such likely future modification and enhancement scenarios raise are: (1) whether the unequal distribution of enhancements is likely to persist or perhaps even widen, and (2) whether the unequal distribution is a significant ethical problem. Although these two questions are distinct questions it is common in the literature that authors who see the inequalities as an ethical problem also think that they are likely to persist or widen, whereas those who think that the inequalities are acceptable or irrelevant also think they are likely to be only temporary (e.g. because of a trickle down from the resource rich to the resource poor).

The view that there is a problem is, for instance expressed by Yoxen in an early contribution to the debate on genetic engineering:

> But at the end of the day, whom will all this activity help? Some people certainly, but, I suspect, they will only be that minority already well supplied with medical goods and services, and the costs of providing cures for the ills of this social stratum will not fall. Too much money is made from putting molecules into people at a profit for very much to change without major convulsions.
>
> That, for me, tarnishes the gleaming image of biotechnology. It is not that most people in health-care companies don't care that their work never touches the lives of millions of people. It is just that they have different priorities: an adequate return on capital invested. Their view of medicine is based on the view that people must first accumulate wealth and then buy health; otherwise there is no deal. The public good may take too long to be a worthwhile business proposition. 'Pure' research is used to fill that gap, but it is being increasingly subject to commercial criteria. (Yoxen 1983, pp. 141–142)

The line of argument employed to show that the inequalities are ethically problematic is usually along Rawlsian lines pointing out that changes in social policy ought to be to the benefit of the 'worst off' (Rawls 1972), and that increasing inequalities by allowing or promoting modification for enhancement is therefore an unacceptable public policy (e.g. Holm 1993).

On the other side of this debate it is claimed that since we allow the resource rich to advance their interests and, for instance 'enhance' their children through non-biological means (better nutrition and private education are often given as examples) there is no reason to single out body modification technologies as particularly bad or problematic and no reason not to allow their use. Or it is argued more directly that the libertarian critique of Rawls put forward by Nozick is decisive and that there is no reason to worry about inequalities along as they have been produced in legitimate ways (Nozick 1974).

Among the participants in the Technolife forum arguments of this type were closely linked to more general ideas about who possesses power in present day western countries (multinationals, media, and government), and to a vision of a more bottom-up individually driven approach to body modification.

### 4.3.4 Controversy 4 – How Far Should We Go?

In the late 1990s and early 2000s a number of disparate pro-enhancement groupings coalesced into what is now known as the Transhumanist movement. The World Transhumanist Association (now Humanity+) was established in 1998 and has since provided the central focus for a movement which describes its guiding idea in the following way:

> Transhumanists view human nature as a work-in-progress, a half-baked beginning that we can learn to remold in desirable ways. Current humanity need not be the endpoint of evolution. Transhumanists hope that by responsible use of science, technology, and other rational means we shall eventually manage to become posthuman, beings with vastly greater capacities than present human beings have. (Bostrom 2013)

The emergence of Transhumanism has re-oriented the enhancement debate, including the debate about body modification. Partly because the Transhumanists categorised everyone who disagreed with them as bio-conservatives and thus, at least in their own mind created a dichotomy in the literature[2]; partly because the overarching nature and ambition of the Transhumanist program moved the focus of attention from individual types of enhancement to the possible effects of multiple technologies converging to produce radical enhancement.

In the current phase of the enhancement debate discussion of justice issues is still prominent, but the arguments do not differ much from those identified above in the early phase. Two new issues became prominent in the late phase: (1) engineered immortality and (2) the 'Singularity'.

One of the enhancements projected by Transhumanists is that the human lifespan can be increased significantly through the use of a combination of technologies. As long as human beings continued to be biological they would still be destructible,

[2] Those identified by Transhumanists as bio-conservatives rarely identify themselves in this way and are not organized in the same way as the Transhumanists are.

so could not achieve "real immortality", but a lifespan of more than 1000 years has been claimed as possible and it has even been claimed that the first person to live to a 1000 is already alive today (de Grey and Rae 2007).

This has sparked a substantial literature concerning whether (1) immortality is desirable and (2) the engineering of immortality creates ethical problems. Many of the ethical problems discussed are variants of the justice discussions outlined above, but new issues are also raised concerning the population effects of engineered mortality for some or all (see for instance Overall 2003, Pijnenburg and Leget 2007).

## 4.4 Technolife Coda

Where does all this leave us? Human beings have always practiced body modification and some of these modifications have probably always been perceived as strange. Not the ones that WE do, in our society, but the ones that THEY do in theirs. So perhaps the current more vociferous controversies merely reflect that we (1) live in societies that are much more diverse than they used to be, and (2) that the 'technical' possibilities for modifying our bodies are ever increasing. But which of the many arguments should we give most weight? If the modification only affects me, then it seems sensible to allow me a wide area of freedom to change my body. But if the change also affect others, justice considerations come into play and the more enhancing a modification is, the more important it becomes that everyone can get access to it. This is where societal judgment and societal action becomes essential. But for that judgment to be legitimate we all have to participate in the discussion and the decision making.

One very prominent feature of the Technolife deliberative exercise on body modification was, that although participants used many of the individual arguments outlined above, many of them did this within a much larger social and technological frame or vision. Questions of liberty and justice were connected to ideas of rapid technological and societal change, and visions of what kind of society would and should develop in the future. This was true of both those who self-identified as Transhumanists and of others. In these visions the possibility to actively influence policy decisions through organisation and information exchange on the internet, the development of non-proprietary technologies and greater transparency were all linked together with the specific arguments about the desirability or not of bodily modification. These linkages were often accomplished by using the internet and its development as a metaphor for, or analogy to specific ways in which enhancements and bodily modifications could occur. Whereas the academic literature on body modification and enhancement is almost exclusively within a biological frame, the Technolife discourse did at least partially move away from that frame to an alternative framing of the internet revolution as the appropriate paradigm for imagining the future of body modification. Importing this alternative framing back into the academic discourse may open the way for new and interesting conceptualisations of nature, liberty and justice and their relation to body modification and its societal embedding.

## References


Beaufort, Inez, Medard Hilhorst, and Søren Holm (eds.). 1996. *In the eye of the beholder – The ethics of appearance changing medical interventions*. Oslo: Scandinavian University Publisher.

Bostrom, Nick. 2013. *Transhumanist values*. http://www.nickbostrom.com/ethics/values.html. Accessed 10 Dec 2013.

Buchanan, Allen. 2011. *Beyond humanity?* Oxford: Oxford University Press.

Daniels, Norman. 1985. *Just health care*. Cambridge: Cambridge University Press.

Daniels, Norman. 2009. Can anyone really be talking about ethically modifying human nature? In *Human enhancement*, ed. Savulescu Julian and Bostrom Nick, 25–42. Oxford: Oxford University Press.

de Grey, Aubrey, and Michael Rae. 2007. *Ending aging: The rejuvenation breakthroughs that could reverse human aging in our lifetime*. New York: Saint Martin's Press.

Farley, Margaret A. 2001. Roman catholic views on research involving human embryonic stem cells. In *The human embryonic stem cell debate*, ed. Holland Suzanne, Lebacqz Karen, and Zoloth Laurie, 113–118. Cambridge, MA: The MIT Press.

Finnis, John. 1980. *Natural law and natural rights*. Oxford: Clarendon.

Harris, John. 2007. *Enhancing evolution*. Princeton: Princeton University Press.

Harris, John. 2009. Enhancements are a moral obligation. In *Human enhancement*, ed. Savulescu Julian and Bostrom Nick, 131–154. Oxford: Oxford University Press.

Holm, Søren. 1993. Genetic engineering and the north–south divide. In *Ethics and biotechnology*, ed. John Harris and Anthony Dyson, 47–63. London: Routledge and Kegan Paul.

Holm, Søren. 2006. The nature of human welfare. In *Future perfect*, ed. Deane-Drummond Celia and Scott Peter Manley, 45–55. London: T & T Clark.

Holm, Søren. 2007. Naturalness and anthropology in modern bioethics, with a special view to trans- and post-humanism. In *Theology and science – Issues for future dialogue*, ed. Helge Kragh Kragh, 17–29. Aarhus: University of Aarhus.

Holm, Søren, and Michael McNamee. 2011. Physical enhancement: What baseline, whose judgment? In *Enhancing human capacities*, ed. Julian Savulescu, Ruud ter Meulen, and Guy Kahane, 291–303. London: Wiley-Blackwell.

Mill, John Stuart. 1978. *On liberty*, ed. E. Rapaport. Indiana: Hackett Publishers.

Murray, Thomas H. 2006. Enhancement. In *The Oxford handbook of bioethics*, ed. Bonnie Steinbock, 491–515. Oxford: Oxford University Press.

Nozick, Robert. 1974. *Anarchy, state, and utopia*. Oxford: Basil Blackwell Ltd.

Oderberg, David S., and Timothy D.J. Chappell (eds.). 2004. *Human values: New essays on ethics and natural law*. London: Palgrave Macmillan.

Overall, Christine. 2003. *Aging death and human longevity: A philosophical inquiry*. Berkeley: University of California Press.

Parens, Erik. 1998. Is better always good? The enhancement project. In *Enhancing human traits*, ed. Parens Erik, 1–29. Washington, DC: Georgetown University Press.

Pijnenburg Martien, A.M., and Carlo Leget. 2007. Who wants to live forever? Three arguments against extending the human lifespan. *Journal of Medical Ethics* 33: 585–587.

Ramsey, Paul. 1970. *Fabricated man: The ethics of genetic control*. New Haven: Yale University Press, New Haven.

Rawls, John. 1972. *A theory of justice*. Oxford: Oxford University Press.

Sandberg, A. (2001). Morphological freedom – why we not just want it, but *need* it. http://www.nada.kth.se/~asa/Texts/MorphologicalFreedom.htm. Accessed 10 Dec 2013.

Silver, Lee. 1998. *Remaking Eden: Designing human life in the new millennium*. London: Weidenfeld & Nicholson.

Strand, Roger, Kjetil Rommetveit, and Silvio Funtowicz. 2011. TECHNOLIFE Deliverable 5.4.2 – Report: TECHNOLIFE ethical frameworks. http://technolife.no/results/content/filelist_4ce40291-cb5f-4ac1-892b-a45e9171cdd4/1336509200705/the_technolife_approach.pdf.

Takala, T. 2004. The (im)morality of (un)naturalness. *Cambridge Quarterly of Healthcare Ethics* 13: 15–19.

Yoxen, Edward. 1983. *The gene business: Who should control biotechnology*. London: Pan Books.

# Chapter 5
# New Options and Identities: Body Enhancement in Science Fiction Narrative

**Miquel Barceló and Louis Lemkow**

**Abstract** Science-fiction has been defined in many ways, but in this chapter priority is given Isaac Asimov's vision which examines human responses to changes in science and technology. In this context the text uses various examples of science-fiction narratives related to the enhancement or modification of the human body for achieving specific objectives (for example, the augmented combat effectiveness of military personnel or the colonisation of new worlds). The possibility of changes or improvements to democracy and governance are analyzed and the implications of body enhancement in this sense are examined critically centered especially on the concept of "cyborg" and the use of all kinds of implants which was very prevalent in the science-fiction of the 1980s in the so-called "cyberpunk" movement.

The most recent speculations concerning the preservation of human minds through computer systems and networks are examined, especially in relation to the social, political and ethical consequences for personal identity in a world where the automatic reproduction not only of objects, but works of art, or also of minds and human identity constructs are possible. The text is completed with a series of conclusions and reflections of a more sociological character. The notion of science-fiction as a tool for learning the future and its capability of providing a kind of preventive strategy in terms of narratives and warnings about undesirable futures is presented and analyzed.

**Keywords** Science fiction literature • Socio-technical imaginaries • Body enhancement • Speculation • Technology • Ethics concers • Ethics for governance

## 5.1 Introduction: Science Fiction and Body Enhancement

It has proved difficult to define the narrative of science fiction. The statement which is attributed to Nietzsche may apply, that "*Only something which has no history is capable of being defined*". The reason for this lack of an all encompassing definition

M. Barceló (✉) • L. Lemkow
Edifici OMEGA – despatx 129 – Campus Nord UPC, Barcelona, Spain
e-mail: blo@essi.upc.edu



A. Delgado (ed.), *Technoscience and Citizenship: Ethics and Governance in the Digital Society*, The International Library of Ethics, Law and Technology 17,
DOI 10.1007/978-3-319-32414-2_5

is simple: changes have occurred throughout the history of science fiction and no definition exists that fits every moment of its history.

The history of science fiction is usually given as beginning almost two centuries ago with the publication of the novel *Frankenstein; or, The Modern Prometheus* (1818) by Mary Shelley. Following this, it has proved difficult to find common categorically defining elements in the narratives of Jules Verne, Herbert G. Wells, Edgar Rice Burroughs, Isaac Asimov, Frederik Pohl, Ursula K. Le Guin, Dan Simmons and Neal Stephenson. However, they all write science fiction. For this reason, some succinct definitions that were practically conceived as *boutades*, complete with an element of surprise, have proved remarkably valid. For example that of Judith Merril: "Science fiction is the literature of disciplined imagination". Or, even better, that of Tom Shippey: Science fiction "is the literature of change, and it changes while you are trying to define it."

In relation to what is of interest to us here, a reflection on the effects of technology on human life and how this poses new challenges in the field of ethics, democracy and governance, the definition, partial as indeed are all the others, offered by the famous author Isaac Asimov may be of use: "Science fiction is that branch of literature that deals with human responses to changes in the level of science and technology".

Whatever it may be, science fiction certainly shares two essential elements that have changed their relative weight and significance over time. The first is what might be called the "*counterfactual conditional*" and relates to the speculative capacity of good science fiction. It consists of asking oneself what would happen if…, and then introducing a hypothesis of something that does not yet exist in reality but can be imagined. Something that is perhaps even within the reach of future developments in science and technology. For example, to ask oneself what would happen if it were possible to travel back in time. If contact could be made with an alien civilization. If it was possible to change the human body and/or the mind to open up new possibilities, and a great long list of other hypotheses. The other element which is perhaps present in all good science fiction, is that known as the "*sense of wonder*", the inevitable surprise of the reader when faced with the new worlds, characters and societies proposed by science fiction. One feature that it shares, in a way, with historical novels or travel books, are descriptions of exotic and unfamiliar realities.

Speculation and wonder are therefore the two constituent traits of the narrative genre of science fiction and are those that shape its wide ranging world of invention and reflection. In good science fiction we find everything we need: speculation about techno-science and its effects, new worlds with all kinds of aliens, new societies and novel ways of organising relationships between individuals within a community, speculative revision of history, countless adventures throughout space and time, and a long almost interminable list of others.

It should be remembered that even the term science fiction itself has been the subject of debate. Isaac Asimov attributed to Robert A. Heinlein the proposal of a new name that explicitly leaves out the term science. It is a question of speaking directly of "speculative fiction" that retains, at least in English, the same acronym: SF. Some authors feel more comfortable with the idea of speculative fiction, such as Damon Knight, the great defender of this new name.

## 5.2 Modifying the Human Body

One of the most frequently repeated aspects of more recent science fiction is the alteration of the human body made possible thanks to new techno-scientific developments. New possibilities open to a world in which humans can complement their intellectual capabilities with electronic implants and/or the digitalization of their bodies and minds. Or one in which the modification of the physical characteristics of the human body by means of science and technology can also occur. Such changes to the human body are a longstanding tradition in science fiction. In his reflection on the possibility of bringing animal behaviour closer to that of humans, Herbert G. Wells was already imagining body modifications (animal in this case) in *The Island of Dr. Moreau* (1896), thanks to developments in the biological sciences. Another example, also from Wells, is *The Invisible Man* (1897), where a new drug makes invisibility possible with all the ethical and social consequences that this triggers. There are times when such changes to, or modification of the human body is not without cost, but reflect what appears to be a desire engineered to obtain certain results. A case in point is the possibility of modifying the human body with a view to its potential survival in environments for which it is not adapted, for example in the conditions on other planets.

Science fiction understood years ago, that in the absence of Martians, if there was to be life on Mars it would be necessary to modify the human being to survive in an ecosystem to which our evolution has not adapted us. In *Man Plus* (1976), Frederik Pohl postulated the use of surgery and new artificial organs to complete that which evolution provided us. To explore and live on Mars, *Homo sapiens* would need to become a new being (the *Man Plus* of the title). A cyborg cosmonaut, half human and half robot with improved lungs for breathing in a rarefied atmosphere, multi-faceted eyes adapted to see in the infrared range, armoured skin, wings with solar batteries incorporated to feed his cybernetic half, and a long list of modifications. A few years later in *Moving Mars* (1993), from the perspective provided by new techno-scientific innovations, Greg Bear imagines that the definitive tool for this transformation might be nanotechnology. Filling the body with "nanobots" that permit its adaptation to the Martian ecology and assist it to survive in Mars' rarefied atmosphere, protect its skin from all types of radiation and a long list of new features.

But technology aside, what stands out most as the result of this transformation that appears at first sight to be merely physical, are really the psychological and ethical consequences that it brings. In *Man Plus*, Pohl analyzes not only these physical changes, but the way Roger Torraway, the "modified" protagonist of the novel, starts to feel like a stranger in his own world. Redesigned and modified to survive on Mars, what Roger really becomes is a monster according to standards on Earths. A monster that day after day, as he is altered by the modifications becomes less adapted to the terrestrial environment, but also to the norms of social life on Earth. Transformed into a loner by the effect of the changes, Roger shows us the human compendium of ethical, sociological and social conflicts experienced by the new "homo tecnologicus". Is such a radical transformation of an individual ethical, even from the perspective of any potential reward in terms of space exploration? The answer is not easy, despite the fact that there have always been individuals who have sacrificed themselves for the species. Is it right for this sacrifice to be so great that it involves the prospect of becoming inhuman, as happens to Roger Torraway of *Man Plus*?

Margaret Atwood counts herself among those authors who prefer to speak of "speculative fiction" as she feels uncomfortable with the label of SF referring only to "hard" science. Atwood has written three novels in the more classical tradition of dystopias, in which classic SF themes are dealt with and in which ethical and governance issues related to the use and abuse of science and technology are confronted. These novels are *The Handmaid's Tale* (1985) and *Oryx and Crake* (2003) which continues in *The Year of the Flood* (2009). The last two relate how even in the hands of "enlightened" scientists, biotechnology can lead to ecological disaster due to the creation of transgenic animals such as the "pigoons" (a pig with human stem cells implanted in them to grow human organs as required) or from the use of human organs banks for transplants. Here the objective of improving the human body is taken to extremes by the mad scientist who is the protagonist of Atwood's universe. Who for example attempts to replace violent human behaviour patterns with peaceful ones. As could have been predicted, the outcomes of these genetic experiments are problematic. Similar outcomes resulted from the modifications in the now classic *Limbo* (1952) by Bernard Wolfe, despite the novel being limited to prosthetic changes by the era in which it was written.

In the same vein and practically in unison, Iain M Banks, Ken Macleod, China Miéville and Charles Stross represent a new generation of science fiction writers, holding progressive left-wing opinions that address, among other things, body enhancement and the topic of governance. The role of the genetically engineered elites in the novels of Banks' Culture series, or Miéville's *Embassytown* (2011) provide their authors with the opportunity to keep a critical eye on intra-species diplomacy, governance and political relations more generally.

## 5.3 Imaging Democracy and Governance

As previously suggested, changes in the human standard exist that might suggest new possibilities that are not merely ethical, but relate to the systems of political representation and governance. To this point examples of changes in science fiction have been discussed that affect physical aspects of the human being, albeit with obvious additional consequences in the ethical and social ambits. But other more "hidden" changes are conceivable that, in spite of everything, would not be without major consequences.

The author Nancy Kress takes this concept as her starting point in the short story *Beggars in Spain* (1993), published in 1991 in *Isaac Asimov's Science Fiction Magazine*. The initial idea was to explore what might happen if some humans did not need to sleep. Then she allowed what might be considered to be the "logical" consequences of this to unfold and the original short story eventually became a trilogy of novels (*Beggars in Spain*, 1993, *Beggars and Choosers*, 1994 and *Beggars Ride*, 1996), resulting in one of the few science fiction narratives that focuses its reflection on themes that feature politics and economics. In *Beggars in Spain*, the wealthy Camden family want something special for their future daughter Leisha: so that she does not have to be subject to the mandatory 8 h of sleep a day during which the human mind recovers from a day of action. To achieve this they get in contact with genetic engineering specialists who will need to modify the genes of the future embryo which will become Leisha so that sleep is not a requirement for her.

This is an example of a modification of the human body that is not apparent on the outside, but is not without major consequences. As Faren Miller says "man has invented Superman and, given human nature, the result is social disaster". Obviously, Leisha Camden is simply among the first of these new human beings, the Sleepless, who modified by genetic engineering to not need to sleep, have greater knowledge and power, as they have more active hours to study, to conduct business, in short, to live. A new social and economic elite has been created, that of the Sleepless who are almost "naturally" in conflict with "normal" humans, the Sleepers. To further complicate things, the Sleepless turn out to have greater longevity (possibly in contradiction to all the rules of biology, but any hypothesis is possible in the unencumbered field of good science fiction narrative). From this combination of insomnia and longevity arise the Superbright sleepless, or Superbrights.

In this case, a new form of class struggle and even new types of social groups arise from a modification of the human body with broad implications that are not just ethical, but also political. The Sleepers are separated off into the "donkeys" (a word which carries a pejorative meaning in English) who manage the system and the "Livers", who, like Herbert G. Wells' Eloi, live a life of leisure maintained by the "donkeys". Nancy Kress herself commented during an interview with Locus: "I use the metaphor of those who sleep and those who do not sleep to refer to the larger group of "haves" and "have-nots"[…]. What do those who have owe to those who do not?"

It is precisely this last question that introduces elements of a profound ethical reflection in the narrative of *Beggars in Spain*. One that is similar to the reaction of the "Baldies", the telepathic mutants in *Mutant* (1953) by Henry Kuttner. In this collection of related stories (a *fix –up*), the new mutants try to help the "normal" humans even despite being rejected by them. Or, a more recent example exists in the X -Men comics and movies of what the group of mutants led by Dr. Xavier try to do when faced with the mutants led by Magneto who are planning to dispense with humans. At the core of *Beggars in Spain* by Nancy Kress lies an issue that is now classic in science fiction, that of the new species (although in this case one that has been "constructed" by human scientific activity) which becomes a curious anthropological speculation. A speculation that does not avoid its political-economic consequences and includes both an emotional debate that is undeniably ethical in origin and a genuinely unusual defence of solidarity.

The success of the first novel led its author to delve deeper into the consequences of what, at first, seemed a simple almost superficial change: the existence of some humans who are not forced to sleep through a third of their lives. This deeper analysis is found in the other two novels of the trilogy: *Beggars and Choosers*, 1994 and *Beggars Ride,* 1996 that once again focus on the political consequences of, as Asimov would put it, the "change in the level of science and technology".

In the second novel in the trilogy, *Beggars and Choosers* (1994), Kress focuses principally on the new society formed by the "donkeys" and the "Livers", in what is a bitter parody of our current democracy. Assuming that the Sleepless have been able to provide the world with boundless energy thus obviating the greatest problem faced in the future by the human species, Kress debates the political feasibility of this new society in which, as they are today, science and technology are central. The questions to be answered by this new social organization are expressed in the classic questions; what does it mean to be human? Who should have control over technology? And how are people affected by science and technology? As Kress writes in the novel, "Technology is Darwinian, it spreads, evolves, adapts and leaves out those who fail to adapt to it."

In the third novel of the trilogy, *Beggars Ride* (1997), Kress introduces a new twist to the political issues and the impact of technological developments on society. The Superbrights (the long-lived almost eternal Sleepless) have offered the whole of society what they call the "Change syringe" which provides immunity against disease. However this only serves to increase the differences between groups, with the "Livers" returning to the state of nomadic tribes, maintained and fed as always by their "donkeys" who now live in protected and isolated enclaves. When the supply of change syringes is interrupted the newborn "Livers" seem doomed to suffer all kinds of diseases. The protagonist, a young "Liver" concerned about the uncertain future of her baby, fights to bring the "Livers" to power and to change things once again. Politics and technology are mere human creations and, ultimately, a trilogy like *Beggars* represents a stunning example of humanism in science fiction. In this sense, it is not surprising that the end of the trilogy is not exactly an unconditional love song to technology. All this, lest we forget, derived from a simple change conceived as an "improvement" to the human body using genetic engineering to mean that human beings simply do not have to sleep.

## 5.4 Modified Bodies, Cyborgs and Implant

In mid May 1999 the existence of a patient who was able to "speak" again following an implant placed in his brain was announced in the press. The new "telepath" was John, a 52 year old labourer who had been left quadriplegic as the result of a brain haemorrhage in January 1998. John learned (but only very poorly) to use his thoughts to control the cursor on a computer system and "talk". Dr. Philip Kennedy was the creator of the technology that allowed John, who had been locked in the prison of his own body, to experience something close to telepathy and act on things by means of thought. The neuro-trophic electrode implanted in John's cortex was an empty glass cone 1.5 mm long and 0.1–0.4 mm in diameter. The cone contained two strands of gold able to conduct a low resistance electrical current. The term "neuro-trophic" refers to organic substances contained in the electrode, which, once inserted, help replenish the surrounding tissue. The process lasted for 3 months after implantation, as the nervous tissue adjacent to the electrode connect to it via dendrites that permit the implant to "feel" the discharges of neighbouring neurons.

It was not an original idea. In 1977, at the University of Utah School of Medicine, a series of electrodes were implanted permanently into the brain of a blind patient, who when appropriately stimulated managed to identify lines transmitted by a television camera. A passive application, from the outside in, which contrasts with the surprising activity, from the inside out, of John with his brand new neuro-trophic electrode. In February 2002, 25 years after the experiment was conducted in Utah, the first "artificial retina" was implanted. It was a kind of microelectronic prosthesis that serves to replace cells damaged by diseases such as retinitis pigmentosa or macular degeneration, diseases that cause blindness or severe visual impairment, affecting over 25 million people worldwide.

In fiction, it was possibly Norman Spinrad who, in *Riding the Torch* (1974) first imagined the possibility of a communication system based on technology activated by thought alone. One that would allow direct communication between human brains. This was the first neurological "implant" recorded in science fiction. Spinrad called it *senso*, a less attractive name than that of *cyberspace,* coined by William Gibson to describe the complex world in the novel *Neuromancer* (1984) which initiated the cyberpunk genre so much in vogue in recent years. Very soon after this, science fiction imagined all kinds of brain implants, including implant-chips able to alter the personality of those who carry them. Such as that imagined by George Alec Effinger in the series that begins with *When Gravity Fails* (1987), which features, among others, nothing less than an removable "James Bond implant", to more than predictable results.

Without a doubt, the most modern version of what constitutes a cybernetic organism or cyborg, results from the union of biology and technology to obtain a functionality superior to that achievable by biology or technology alone. It should be remembered that the need to connect the human brain to a technological system had already been imagined in the science fiction of the 1930s. Precisely at a time when it was required to handle very complex tasks, for example, to control the traffic in a big city of the future. An incontrovertible demonstration of the lack of confidence it had in what later, in 1956, would become known as artificial intelligence. Such brains, connected to technological devices from the thirties, would very soon progress to directing other complex technological systems, for example spaceships. Possibly the pioneer in this was Anne McCaffrey with *The Ship Who Sang* (1961). The ship which did sing was actually a girl, Helva, whose deeply defective body enclosed a viable mind. Refusing the option of euthanasia, Helva's parents decide to accept that the mind of the child be trained and programmed to become the entity that controls and directs a new body of titanium, an interstellar scout ship and thus Helva becomes a human without a biological body, a virtually immortal cyborg.

There are other examples of implants, even in the cinema. In this case it is necessary to recall a bad movie, *Johnny Mnemonic*, directed by Robert Longo in 1995 and starring Keanu Reeves, in which the protagonist has an implant that allows him to store data becoming a "human transporter" able to transport this data in a highly secretive manner.

## 5.5 Emerging Identities and Worlds

Unfortunately the *cyberpunk* genre has always appeared to be rather too timid to imagine the potential social impact of large-capacity computer technologies, artificial intelligence, global networks such as the Internet, and a long, long list of others. Not even William Gibson or Bruce Sterling, without a doubt the best authors of the genre of *cyberpunk* that has been so well promoted by its publishers, have adequately exploited this novel speculative vein of the worlds of computing and their new possibilities.

But fortunately things are slowly changing and, leaving aside those commercial factions that have little or nothing to contribute; genuine science fiction authors can be found who are not afraid to let their imaginations run wild in new digital worlds. Some have labelled them *post-cyberpunk* and, quite possibly, Greg Egan and Neal Stephenson count among their greatest champions.

Perhaps the most characteristic of them all is the Australian Greg Egan, one of the few science fiction authors to have a thorough understanding of current computer technology. His novels are always respectful of scientific reality and technological knowhow and include interesting and risky speculation. In P*ermutation City* (1994), Egan imagines that, in the middle of the twenty first century, it will be possible to scan the human mind and store it in a computer as a "Copy". These copies are able to control the virtual reality environment in which they exist, and live a life completely analogous to the life that we know, existing in their own way in a virtual world which is in every way a simulation of our own.

The first question springs to mind immediately and is of a philosophical nature: where does the identity reside? The simultaneous existence of a human being and a copy (or several copies) poses this question in a particularly aggressive manner. We recall that Egan introduces the so called *Dust Theory* almost in passing, according to which human consciousness (or at least that of the copies) is not localized and, like dust, is distributed in space and time. Existing essentially as a structure or pattern and not in a concrete location. In any case, the Copies are an obvious way to overcome the limited duration of the human lifespan. In the novel, the wealthiest humans store themselves as Copies just before their death, in one last pursuit of much sought after immortality. And with success: life as Copy satisfies all their requirements. It is a final state. And it seems durable. The supposed immortality of the Copies has its limit: it is threatened by the possible and perhaps inevitable disconnection of the computers where the complex structure that constitutes the Copy and its setting resides. The computer system in which they exist depends on the real world for the energy required to feed it.

In the novel a select group of Copies in possession of the greatest riches, is offered the possibility of eternal life in a self-replicating expanding cellular automaton, the future *Permutation City* of the title of the novel. It is a strange idea, but one that responds to the scientific speculations already conceived by Alain Turing and John von Neumann in the 1940s and 1950s. Greg Egan allows his final version to exist only in the imagination: the cellular automaton TVC (Turing, von Neumann and Chiang) that appears in the novel as the result of the work of a certain Chiang in 2010. A copy of Autoverse is also stored in the same cellular automaton. Autoverse is a simulator that contains a simplified set of the laws of physics and chemistry and ultimately configures a simulated digital universe in which, after experimentation with bacterial life forms, resulted in the development of a complete alternative evolution to that of our universe, including intelligent life.

The novel's unique *tour de force* lies in the fact that the simulation of our universe in the world of Copies and the new simplified Autoverse based on its laws, coexist in the same TVC cellular automaton. In case this novel, already more absorbing than most, lacked complexity, the question is posed whether one of these sets of laws, one of these universes will ultimately prevail over the other in the cellular automaton where both coexist.

It is a curious speculation that takes us back to the meaning of the laws of nature and the underlying structure of the universe. This is true speculation, and is a genuine intellectual luxury compared to the poverty of many films such as Johnny Mnemonic, which seem to be the most that, in film, the unimaginative cyberpunk genre that surrounded us has managed to produce. *Permutation City* merits its various major awards for science fiction: the John Campbell Memorial and, logically, the Australian Ditmar Award. But Egan, who is regarded by the expert Pedro Jorge Romero to be "an Australian lost in metaphysics", is not the only one who has speculated on an alternative universe of digital and digitized life. There are other examples, albeit simpler and closer to everyday reality.

In *The Terminal Experiment* (1995), which has also received awards, the Canadian Robert J. Sawyer imagined a curious experiment in which a new life form was created in a digital universe. Sawyer makes use of one of the best narrative formulas of modern science fiction: novels that owe much to normal characters embroiled in a mysterious plot that is brilliantly resolved with techniques common in the best thrillers. But in the case of Sawyer, this time the theme is rigorous, well documented science fiction. The science is interesting but always complemented by an intriguing reflection on moral issues and the inevitably subjective nature of cultural and ethical behaviour. At a time in which techno-science and its achievements rapidly modify and alter the global living conditions across the whole world, it is worth questioning the morality and the ethical component of the activity of scientists and engineers and the ultimate consequences of their intellectual creations and work. And this seems to be the great specialty of Robert J. Sawyer, who appears to enjoy a superior speculative capacity and an explanatory ease for the dissemination of science, reminiscent of Asimov at his best.

If Egan's writing is deep, metaphysical and somewhat cryptic, Sawyer's is extremely realistic, transparent and even educational. In the case of *The Terminal Experiment* (which probably should have been given the title "The Final Experiment" for its publication in Spanish), it all started with the *Mid-December* 1994 edition of *Analog, Science Fiction/Science Fact* in which appeared the first part of a serial by Robert J. Sawyer. At that time it was called *Hobson's Choice*. After its publication in March 1995 in book form it became *The Terminal Experiment*, and was destined to become a finalist in the 1996 Hugo Award and the clear winner of the 1995 Nebula prize. In the same way that previously it had won the Canadian Aurora award (not to discourage patriotism) and the annual Compuserve Science Fiction Forum Homer award.

Suffice to say that *The Terminal Experiment* presents Dr. Hobson, a "normal" character (if that makes sense?) faced with a problem which could also be said to be quite "normal", that of his wife's infidelity. In the context of techno-science that may soon also become normal. After discovering what may be the electrical trace of the soul (remember that this is science fiction). Dr. Hobson attempts to study novel concepts of life and death. He does this by means of computer simulations of his own brain, discovering that things might appear normal, but are never as simple as they seem.

In reality, Dr. Hobson has created a monster. Or in fact, he has created three. To test his theories about immortality and the possible existence of life after death, Hobson manages to create three computer simulations of his own personality. With the first, from which all references to physical existence have been removed, he attempts to study what a possible "ethereal" life would be like after physical death. With the second, from which any reference to aging and death have been eliminated, Hobson intended to study what it was like to live with an intrinsic feeling of immortality. The third unaltered simulation, serves as the control reference for the experiment. The three simulations escape from Hobson's control, flee the computer where they are confined and lodge in the global computer network to live out their own lives. And one of them happens to be a murderer. A murderer who actually carries out crimes that the mind of the biological Hobson may have imagined and even desired to have committed himself.

That's the idea, it is basically a straightforward murder mystery (the objective being to find the murderer), with uncomplicated motivations, and simple yet interesting approaches to the reasons why certain things happen. *The Terminal Experiment* justifies perfectly why it received its awards. The speculations of both Egan and Sawyer bring us closer to a new reality which (maybe, just maybe) technology might in a more or less distant future make possible: if we could live in the virtual world of the internet, in a virtual universe, how would we recognize our reality in this virtual universe? Bishop Berkeley would be happy.

## 5.6 Identity in the World of Technical Reproducibility

By 1934, Walter Benjamin had already predicted what would happen to intellectual property (or, more accurately in his case, what would happen in an art market characterized by the uniqueness of a work of art or in the case of lithographs a limited number of works) when a technique permitted the reproduction of works of art with practically no loss of pictorial content. He outlined this prediction in a paper entitled "The Work of Art in the Age of Mechanical Reproduction" (1934). Something similar can be imagined in relation to personal identity, when science fiction imagines that "copies" of human beings can be created. And we are no longer speaking of those digital "copies" that can only live in the internet or in the digital world, such as those in *Permutation city* or *The Terminal Experiment*. They refer to true copies with a human appearance and the same potential as humans. Since the cloning of Dolly the sheep was announced in February 1997, the idea of potential human cloning has generated various debates, and it must be said that several science fiction short stories and novels take their starting point very much from Dolly.

Some authors, such as David Brin have not stopped here. Without specifically addressing the idea of clones, they have analyzed a new idea, similar but not exactly the same: a "golem" like species of beings made of clay fabricated in new copier-ovens that, 50 years into the future, are within everyone's reach and form a new type of society. These "ditto" or *Kiln People* (2003) have a limited life-span of one day, and are not biological beings like humans, but, as their name implies are made of clay. As would be expected, they lack legal and social rights and even come in different colours according to their function. They are given the less interesting and most dangerous occupations, all those that human beings do not wish to do. At the end of their existence, if possible, the ditto "download" into their original character, known as the "archi" from archetype, the memories collected on that day. A day that begins with the imprint of the memories of their human original, or archetype, on their own minds.

Building on this idea, Brin imagines that in the near future everyone is able to make these dittos, these "clay people" and, in a coherent and well analyzed way, he describes to us a complex society where humans end up living parallel lives. Their own and that of the dittos that have been able to, at the end of their day of existence, download their memories into the original. What really interests Brin is social analysis. He describes an example of this new society to us on his website: "As a citizen of a near future, you have duplicated yourself zillions of times and you see it as something that is quite normal, being sometimes the original and sometimes the copy. You live life in parallel, sending expensive "study golems" to the library, while other cheaper models clean your house and your real body exercises in the gym. Two thirds of the population of Earth are temporary beings made of clay. People seem to have adapted to this new kind of life, until…" and this "until…" triggers a crisis in this new society of the future where duplicate beings are within everyone's reach.

Not forgetting that all social reflection in the form of science fiction novels should be entertaining, Brin has used the form of a narrative detective action, from the hard -boiled genre, to demonstrate the complexities of a society in which exist a curious version of the cinematographic "replicants" from Blade Runner. The dittos are derived from human beings, "produced" by humans, have similar "features" to human beings, but have an expiry date, as if they were yogurt. The novel chronicles the adventures of the private detective Albert Morris and his multiple clay duplicates in this new society. In Dittoburg they are making pirate copies of a famous courtesan, Gineen Wammaker, and Morris has to stop this. A job that does not seem too difficult, but it leads him to discover an intricate web of conspiracies in this society of the future where the ditto lack any rights or any respect.

Returning to the reference to the "replicants" in *Blade Runner* or, if you will, to the "androids" that Philip K. Dick spoke of in his novel *Do Androids Dream of Electric Sheep*? (1967), the fact is that the years do not appear to have passed in vain for science fiction. Dick's novel contains the very simple model based on two central points:

(a) Who is the machine (the android) and who is the human? And
(b) the sensation that androids have of being human, who should be granted full legal and social rights, even though those rights will not be recognized.

Here, in a way this is repeated, but in the midst of many other considerations. Science fiction has grown up and when it analyzes a possible future, some authors at least (and Brin is certainly one of them) contemplate society as the complex entity it really is and analyze it from diverse points of view.

In *Kiln People*, there are aspects of criticism of slavery in the way that dittos are used (and abused) and there is a reflection about what it means to be human and the rights a human or a simulacrum should be granted. Brin does not spare any effort in imagining the many different ways of using the new technology of the copier-ovens (social imagination appears to be endless) and a long list of other considerations.

In the same vein the book *Mindscan* (2005), by Robert J. Sawyer, mentioned previously, deals very clearly with the legal, personal, human consequences that would result from a possible discharge of a human mind into an android body. This is very similar to the technological theme underlying the short story *Identity Theft* with which Sawyer won (in 2004 for the umpteenth time) the UPC Science Fiction Award. In this case, it is a detective story with a protagonist that emulates classic characters like Sam Spade or Philip Marlowe, this time in a sci-fi environment: on Mars in the future. The new technology, here called the "Transfer" (a human mind transferred to a new artificial body made with biological and robotic techniques), causes serious problems when the private detective Lomax Alexander tries to discover the whereabouts of a missing husband.

In *Mindscan*, the idea is the same, but the technology has changed name and is now called "mindscan" (in reality it is a quantum mind scanner). With it, Jake Sullivan manages to cheat death. Threatened by a cruel and fatal circulatory disease, he manages to copy his consciousness into a new android body (referred to as the Jack mindscan) that is to provide him with immortality. Other very elderly billionaires do the same. Thanks to the new "mindscan" process by Inmortex, the "originals" are retired with their obsolete and outdated biological bodies to a luxurious centre on the other side of the moon, giving up their place on Earth to their mindscan copies. Not everything, however, turns out as expected. Although the new Jake mindscan finds love that the disease previously prevented him from having, his new partner, a popular mindscan writer, must face a groundbreaking trial because her son does not want to renounce his inheritance and refuses to consider the new Mindscan copy human. In the mean time, on the moon, the real biological Jake has cured his circulatory disease thanks to new developments in medical nanotechnology and decides to return to Earth. Who is who? Who retains the rights of the individual: the biological original or the Mindscan to whom they have been ceded? What does it mean to be human? In short, *Mindscan* becomes a surprising personal and legal drama and, basically an interesting speculation about a not too distant future in which, due to technology, not even one's own identity is assured.

## 5.7 Living in a Digital World

*Rainbows End* (2006) by Vernor Vinge, clearly falls within the theme of postulating, for a not so distant future, no longer for artificial intelligence, but for a true revolution in the world of human intelligence through the open collaboration of many human minds with the possibilities offered by the new technology. This is how it was seen by no less than Thomas W. Malone, who is the *Patrick J. McGovern Professor* at the Massachusetts Institute of Technology's Sloan School of Management where he also directs the *Center for Coordination Science* at the institute. Malone is quoted as saying the following about Vernor Vinge's latest novel: "*Vernor Vinge has done it again: he has predicted the human implications of new technologies much earlier than any other. In* True Names *it was cyberspace. In* Rainbows End, *among other things, the superhuman intelligence that can result from intelligent electronic connection between a vast number of human beings. It is not just a good story, but one that resulted in my neurons coming face to face with unexpectedly new ideas every few pages.*"

All this comes from the new interface with which humans use the powerful info-technologies of the near future. One such interface is Epiphany, which uses "wearable" computer elements and falls squarely within what is known as "augmented reality", a field of modern research in computer science that involves the combining of the real world with computer generated data. A "new" global reality in which humans do not only interact with real elements, but also with computer-generated virtual elements. This can be done at any time as it is no longer dependent on windows, icons or mice, as is the case with the current dominant computer interface. Epiphany's special contact lenses and "wearable" computer can replace what the eye would see naturally with a visual construction created by computer graphics that substitute and/or supplement reality.

As the perfect complement, "haptic" feedback makes direct participation possible in this new reality. "Haptic" technology is also currently being researched and refers to the technology that uses the sense of touch as the interface with the user by applying forces, vibrations and/or movement that the user perceives with their sense of touch as preceding from "real" reality, so to speak. It is no longer about individual implants, but rather technology that might be labelled "social" and which obliges us to redefine the role of the individual in a society of alternative and virtual selves that, in this case, respond to new developments and the possibilities opened up by new technology.

Evidently, all this technological paraphernalia exists as a baseline reference, omnipresent but not invasive and, as is usually the case in a brilliant storyteller such as Vernor Vinge, the interest in his novels lies not only their ideas (which are interesting and plentiful) but in the plot itself, the characters and the difficulties which they must face and overcome. To achieve this, Vernor Vinge uses the figure of a famous aged poet Robert Gu who has been afflicted for many years with Alzheimers disease, but for whom the medicine available in the near future will be able to offer recovery and rejuvenation bringing him face to face with a similar but insidiously different world to that he remembers. We discover the somewhat abusive personality of the protagonist when faced with his memories, undergoing what may be a final crisis of creativity, surprising in its intelligence and humanity, as a result of the new composite reality that the world has become. An utterly spectacular genuine change owed above all to the collaboration of millions of human intelligences on the internet to which "Epiphany", the new computer interface based on "wearable" computer technology which is to replace the now ubiquitous system of windows, icons and mice is accessed.

The speculative part of R*ainbows End* comes from a variation of the fictional world that Vernor Vinge established in the short story "Fast Times at Fairmont High" (2002), which also won the Hugo Award in the short story category, and to which other elements are also added from the later story: "Synthetic Serendipity" (2004). It evidently owes much to current research into ubiquitous computing, wearable computer technology, haptic interfaces and everything that is on the crest of the wave of the latest research in the field of info-technologies. An element of this is made directly explicit in the novel's dedication: "For Internet based cognitive tools that are changing our lives: Wikipedia, Google, eBay and the like, now and in the future".

In an interview given to the Argentinean magazine "Ñ", conducted by Andres Hax, Vernor Vinge also introduced a political element into his reflections and, ultimately, into the new perspective that his hypothesis takes with regard to "technological singularity" *"I believe that there is a high probability that the internet, i.e. computers, databases and associated humans, are a new phenomenon. An unprecedented one in history. If you simply do the figures, you will see that there are hundreds of millions of educated people today who are connected on the internet. And within this huge mass there are probably a million people who are smarter than all the people who in the past ran the world. In addition, you can find thousands of people who happen to be experts in anything you want to know about, on any subject, no matter how insignificant or unexplored the issue. All these people are working on their own, free from central control, but they are contributing to the fact that for the first time in history, you don't have to be extremely rich or holy to understand and be connected to the* Big Picture *of what is happening in the world.*

*In the last 200 years or so, people have been approaching this slowly. But now it is such an intense phenomenon that the hope exists that it will even prove enough to minimize the influence of governments on humanity. It doesn't mean that governments will disappear. Just that they will basically be tolerated as infrastructure providers rather than commanders of our faith".*

This is it seems, the idea that moves Vernor Vinge to speculate about the new interface, Epiphany, which generates new possibilities of interaction between humans (who remain such, as demonstrated by the personality of the protagonist, the poet Robert Gu), and with them, new social forms and, evidently policies as well. Perhaps for this reason, Vernor Vinge himself feels it necessary to justify this speculative choice that, as is always the case, is usually the best background to this brilliantly speculative narrative fiction: "*Science fiction does two things. The most important thing is that it stimulates young people to think about science and to enter careers in technology. The other thing it does, somewhat more grandiose sounding, is that the stories of science fiction writers are for the body politic of humanity what dreaming is for the human individual. And you know, of course, that dreams in general have no predictive value. But from time to time they generate very good ideas that otherwise would not have occurred to you.*" With Nancy Kress's permission.

## 5.8 By Way of Conclusion: The Role of Science Fiction in Learning About the Future

The attention that much of today's science fiction pays to the near future and to current key technologies (biotechnologies, info-technologies, etc.) strengthens its role as a tool for learning to live in the near future. A future that we only know (due to the speed of the change promoted by modern science and technology) will be different from the past and the present in which we have lived. Ogburn's ideas established nearly a century ago and popularised in *Future Shock* (1970) by the American writer Alvin Toffler, lead us to reflect on the rate of change in a modern culture dominated by the effects of techno-science and subject to its exceptional ability to transform.

The reader of science fiction, an enthusiast and in a certain way specialist in imagining and confronting distinct futures created by techno-science, acquires specialist learning which makes them better suited to living in the future that awaits us just around the corner. It is true that the future will not be, by any means, exactly as described in science fiction. The mere act of thinking about the relativity of the present and in the possible alternatives that may arise in the future may well be, for readers of science fiction, good training to survive in the concrete future, that ultimately, may turn out to be true. Of the many examples of this, the most obvious recent example is that found in the analysis of social reaction to the possibility of genetic engineering and, specifically, to the viability of human cloning. For the majority of society, this is an idea that did not receive any attention until, in February 1997 with *Dolly* the sheep; the cloning of the first mammal was achieved.

However, readers of science fiction have long contemplated this possibility. Since 1932 with the publication of Aldous Huxley's *Brave New World*, (inspired by the work and certain speculations of the biologist J. B. S. Haldane, 1892–1964) and above all since the decade of the 1970s which saw science fiction reflecting abundantly on the social, psychological and even military aspects of possible human cloning and genetic engineering. In stories like "Nine Lives" (1968) by Ursula K. Le Guin, and in novels such as *Where Late the Sweet Birds Sang* (1976) by Kate Wilhelm, *The Ophiuchi Hotline* (1977) by John Varley, or the remarkable and visionary *Cyteen* (1988) by C. J. Cherryh, science fiction readers have learned to imagine what the "human response to changes in the level of science and technology" might be in relation to the cloning of human beings. This long before *Dolly* the sheep, providing an accurate indication of the ability of science fiction to get closer to what we might call learning about the future.

Science fiction remains, therefore, one of the best tools for this learning. When the author of the narrative is sufficiently skilled, this imagined learning becomes somehow 'lived'. It becomes a form of "practice" and makes possible the learning required to live in the distinct future that awaits us. If good science fiction describes possible futures, a valid approach would be to ask why it does this and how it could be made use of in the present. The examples given in this chapter bring new realities of social life closer, realities that thanks to the growing use of science and technology are no less real and/or possible despite being imaginary and in their early stages.

From the elementary modification of the human body (*Man Plus* by Pohl, *Moving Mars* by Bear and *Beggars in Spain* by Kress) that the science and technology of an immediate future can provide, society is able to make the leap to new possibilities of social relationship similar to the extreme examples such as those presented in *Permutation City* by Egan, *Kiln People* by Brin and *The Terminal Experiment* and *Mindscan* by Sawyer or, much closer to home in *Rainbow's End* by Vinge. All are bold speculations that act as a warning and, at the same time as possible preventive prophecies that might avoid the occurrence of undesirable extremes such as those in classics like *A Brave New World* by Huxley, or Orwell's 1984 and Wolfe's *Limbo*.

In this way science fiction becomes an indispensable element and tool to investigate the possible society of the future and the relationships that individuals might establish through the use of new techno-sciences, either as preventive prophecy to avoid undesirable futures or as a way to learn about the unknown future that awaits us. In a sense it follows what H. G. Wells, the true founding father of the genre, advocated in a talk he gave to the British Sociological Society in 1906. The father of modern science fiction recommended that sociology adopt its own "appropriate and distinctive approach" for creating utopias and comprehensively evaluating them. This game of imagining futures (utopian or not) and also of warning of the hazards implicit in certain modern trends, is one of the most enriching aspects of the speculation of science fiction.

## References


Asimov, I. 1981. *Asimov on science fiction*. New York: Doubleday.
Atwood, M. 2003. *Oryx and Crake*. New York: Doubleday.
Barceló, M. 1990. *Ciencia ficción: guía de lectura*. Barcelona: Ediciones B.
Bear, G. 1993. *Moving Mars*. New York: Tor Books.
Brin, D. 2003. *Kiln people*. New York: Tor Books.
Cherryh, C.J. 1988. *Cyteen*. New York: Dawn Books.
Egan, G. 1994. *Permutation city*. London: Millennium: An Orion Book.
Huxley, A. 1932. *Brave new world*. London: Chatto & Windows.
Kress, N. 1993. *Beggars in Spain*. New York: Harper Collins.
Kress, N. 1994. *Beggars and choosers*. New York: Tor Books.
Kress, N. 1997. *Beggars ride*. New York: Tor Books.
Le Guin, U.K. 1968. *Nine lifes* in Playboy Magazine, and in *The wind's twelve quarters* (1975). New York: Harper & Row.
Miéville, C. 2011. *Embassytown*. New York: Del Rey Books.
Orwell, G. 1949. *1984*. London: Secker & Warburg.
Pohl, F. 1976. *Man plus*. New York: Random House.
Sawyer, R.J. 1995. *The terminal experiment*. New York: Harper Prism.
Sawyer, R.J. 2005. *Mindscan*. New York: Tor Books.
Toffler, A. 1970. *Future shock*. New York: Random House.
Varley, J. 1977. *The Ophiuchi hotline*. London: Gollanz.
Vinge, V. 2006. *Rainbows end*. New York: Tor Books.
Wilhelm, K. 1976. *Where late the sweet birds sang*. New York: Harper & Row.
Wolfe, B. 1952. *Limbo*. New York: Ace Books (1963).

# Part III
# Governing Citizen's Movements

## On Biometrics, Securitization and Surveillance

# Part II

# Chapter 6
# The *Fifth Freedom* and the Burden of Executive Power

**Kristrún Gunnarsdóttir**

**Abstract** The deployment of advanced ICTs and biometry in mass-surveillance and border control is part of the *securitization agenda* that emerged in the early 2000s. This agenda has been particularly instrumental in cultivating migration anxieties and framing the problem of *threat* as an imperative to identify those who are dangerous to public safety. As well founded as that may be, it masks the pivotal role ICTs have in the supervision and administration of industries and markets. ICTs, including biometrics, are essential to supervise all four freedoms of movement in European market integration, i.e., of goods, services, capital and persons. They are essential to the EU-US trade and investment relationship which is increasingly underpinned by cross-border data flows, and the most significant such relationship globally. Drawing on mobilities research, this chapter explores how the mobilities of materials, commodities, markets and labour are simultaneously constrained and facilitated in transnational development, including the obligation in Europe to protect ***the fifth freedom*** of movement, that of data. While a recent European regulation aims to better protect personal data in these flows, the narratives of threat and emergency call for immediate action, whereby any data that can be intercepted can also be gathered for investigative purposes on the basis of *exceptional circumstance*. The securitization agenda finds its practical utility in the hands of executive powers who are largely avoiding the legislature and the judiciary. There is no evidence that authorities catch terrorists and criminals *because* of advanced ICTs and biometry in border control and data intercept. The practical utility lies in the ability to target and investigate any individual and a whole range of political opposition and activism which has very little if anything to do with threat and emergency. Publics have no meaningful defence apart from self-censorship. They have no meaningful way of objecting to exceptional circumstances in which illiberal practices are legitimized, and neither does the legislature and the judiciary unless the checks on executive powers are adequately reined in.

K. Gunnarsdóttir (✉)
Sociology Department, University of Surrey, Guildford GU2 7XH, UK

Lancaster University, Lancaster LA1 4YN, UK
e-mail: k.gunnarsdottir@surrey.ac.uk



A. Delgado (ed.), *Technoscience and Citizenship: Ethics and Governance in the Digital Society*, The International Library of Ethics, Law and Technology 17,
DOI 10.1007/978-3-319-32414-2_6

**Keywords** Fifth freedom • Free movement of data • Freedom of expression • Biometry • Surveillance • Securitization • Data protection • Data-driven markets • Economic growth • Economic leadership • Executive power

## 6.1 Instrumentality and Practical Utility[1]

*Security* is the watchword with which to refer to public safety and safeguarding first-world economies, markets, democracy and *our* mobile lives. The growth of industries and markets since the second world war coincides with the freedoms of movement that are protected across the Single Market of the European Economic Area (EEA), and in market relations with the United States and beyond. The sustainability of such freedoms is increasingly at issue however. Evidence of political dissidence and hostility toward the current socio-economic order and the leadership of first-world democracies and corporate enterprise, give rise to insecurities for which securitization has become the all-round remedy.

This chapter addresses a set of issues that tie together securitization objectives in reference to industries and markets, economic leadership and the mobilities afforded by European integration and associated transnational developments. The EEA protects the so-called Four Freedoms: the *free movement of goods, services, capital and persons* (European Parliament Fact Sheets 2000). Among other things, it supports the abolition of customs duties on industrial goods, the free movement of financial services, telecommunication, information, media services, transport and energy, the freedom of ownership and right to take up residence and employment in any of the EEA member states without visa formalities. Accordingly, the supervision of cross-border mobilities in Europe falls on shared regulatory provisions, including the bureaucratic and technical oversight of the EEA and Schengen agreements.[2]

As regards relations beyond the EEA however, European states have tacitly accepted the leadership of the United States in economic, political and military affairs, and relations outside of Europe and the US are marked by a history of colonialism and exploitation. Calls for democracy on behalf of some of those *others* comes on the back of coercive social, economic and market practices, instigated and policed by the US and former European empires (e.g. Dalacoura 2005; Carothers 2003). While conditions like that give rise to questions of adequacy with respect to *our leadership*, and to concerns about economic, social and territorial cohesion,

[1] The arguments presented in this chapter draw on sociological and policy research for two FP7-funded projects, Technolife, No: FP7-230381 (http://neicts.lancs.ac.uk/old/technolife.htm) and ICTethics, No: FP7-230368 (http://neicts.lancs.ac.uk/ictethics.htm). It is supported by two case studies within these projects, titled – *ICT for Human Security* and *Biometric Technologies* – but also on participation in expert meetings and workshops organised by the FP7-funded projects RISE, HIDE, and the BEST Network.

[2] The Schengen agreement designates a border-less geographical area within Europe with border controls for travelers crossing the Schengen frontier borders but, ideally, not the internal borders.

transnational partnerships are adapting to signs of change. But, as the official argument goes, first-world leadership will have to remake itself in an era of new economic powers, and "[t]hat begins with our economic leadership" (Obama 2011).[3] With that in mind, the most significant trade and investment relationship globally is the one between European states and the US, a relationship increasingly underpinned by cross-border flows of data (Meltzer 2014).

This chapter takes as a point of departure that the proliferation of advanced ICTs in transnational mobility control and surveillance, has very publicly served a preoccupation with imminent threat of illicit migration and acts of crime and terror. It argues that such preoccupation, regardless of how well founded it may be, masks the pivotal role of ICTs in the practical supervision and oversight of transnational markets, industry and finance. The following sections discuss the question of cohesion in transnational development, the underpinnings of markets in data flows and ICT-based innovation, and how to move away from instrumentalizing the securitization agenda to a point where publics have no meaningful way of challenging it. To put it rather crudely, the agenda to-date has been dominated by a no-win scenario: would 'you' not give up some personal data and privacy so authorities can catch terrorists and keep you safe? This particular framing is not systematically challenged – of course you would – while it holds within the self-evidence of interoperable and integrated ICT systems, of biometric identification, visa registries and related applications (Amoore 2006). Ethicists and legal scholars have taken more or less at face value the focus here in governance on *striking the right balance* between the freedoms of individuals and providing them with security. Among other things, their work attempts to clarify what is at stake if a right to privacy and self-determination is taken away (see Laas-Mikko and Sutrop, Chap. 7, this volume). New privacy-enhancing technologies have seen light, new data processing principles, and attempts are made to improve on existing legal frameworks in Europe, e.g., with the new General Data Protection Regulation (GDPR) (European Parliament and the Council of the European Union 2016). However, widely researched and debated concerns over first-world liberties and citizen rights, draw attention away from the role of the securitization agenda in protecting the free movement of data. Against concerns that rights and liberties are at risk and in need of protection from the very methods of delivering security, the free movement of data is the *versatile materiality* of economic competitiveness and growth. Productivity and cohesion depend on it along with the pursuits of other freedoms of movement with which transnational industries and markets progress.

Taking these considerations into account, the question remains how to re-frame the securitization agenda to foreground the issues that are obscured by the no-win scenario and the metaphor of striking a balance. Firstly, there is nothing to prove that authorities catch terrorists and criminals *because* of biometric registries and cutting-edge ICTs. There is nothing to suggest categorical change in a long history of record keeping and investigative techniques for operational and administrative

[3] US President Obama speaking to both houses of Parliament in the United Kingdom, 25 May 2011.

purposes of states and constabularies, who have been at times vulnerable to dictatorial tendencies. Advancing ICT-based surveillance and record-keeping, and implementing biometry for identification, is largely a matter of efficiency, convenience and service quality (see Rommetveit, Chap. 8, this volume). Secondly, when authorities identify threat and justify surveillance in the name of security, they appear to be targeting political activism against war and injustice, members of Occupy (Poitras 2014), and those who mobilise against various scientific, technological and other societal innovations (e.g. Welsh and Wynne, 2013). In reference to *exceptional circumstance*, framed in militarist terms as *threat* or as criminal investigation, authorities will simply decide what they need and what to do, and involve publics by command, not consent and consultation (see Agamben 2005 on state of exception). In the United States, authorities are known to subpoena the online service industries and other commerce for the *personal* data they process, i.e., to gain insights into who the culprits are or these *publics-as-threat* on their watch lists—to infiltrate their ranks and neutralise their impact. The result is a growing number of self-censoring publics while we also learn from the Snowden revelations that authorities in European states have been depending upon access to precisely the same sorts of insights with assistance from the United States (e.g. Harding 2014). This *practical utility* of the securitization agenda has been shrouded in secrecy while the agenda is still *instrumental* in amplifying visions of vulnerable – but irrelevant – publics, apparently ignorant of the security threats surrounding them.

## 6.2 Mobilities: Associations and Relations

Transnational development in tourism, migration, markets and mobile employment are among the many study topics found in the academic discourse on contemporary mobilities. Mobilities research is focussed on the technological and material conditions of persons, objects and ideas on the move and in stasis. It aims to clarify the political, cultural and socio-economic relevance of mobility or the lack thereof (e.g. Gill et al. 2011; Scuzzarello and Kinnvall 2013; Jordan and Brown 2007; Stephenson 2006; Tyler 2006; Urry 2007). We learn how mobilities feature in personal, occupational and organizational lives, including the many "[i]ssues of movement, of too little movement or too much, or of the wrong sort or at the wrong time" (Sheller and Urry 2006: 208). Not only do persons and things flow through transits but also through cyberspace by way of representation and re-creation of memories, imaginaries and governance, using transportable media. In short, mobilities are deployed and consumed in the course of achieving what are considered normal commitments and obligations (Shove 2002). They satisfy a need to sustain proximity (Urry 2002), a *will to connection* (Sheller and Urry 2006), and they require constant management.

The EEA and Schengen agreements are only two examples of a whole complex of relations and associations Europeans are committed to across the region, to align and mobilize social, economic and market interests (Fig. 6.1). The countries of

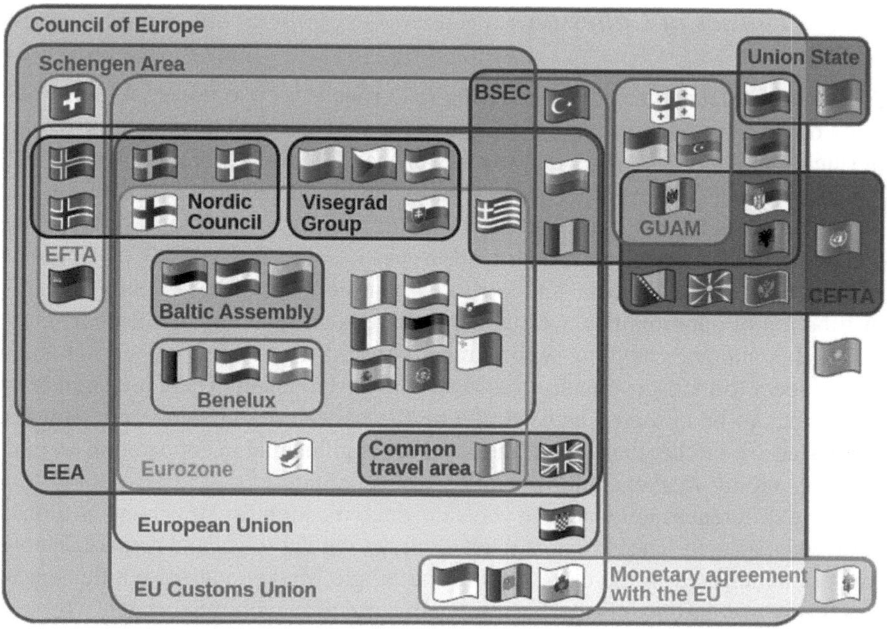


**Fig. 6.1** A diagram showing supranational relations and associations within the boundaries of the Council of Europe, plus Vatican City, Belarus, Kazakhstan and Kosovo. Associations across Europe change over time and so have the Euler diagrams in this and related Wikipedia entries, http://en.wikipedia.org/wiki/European_Economic_Area (Sept 2016)

Europe also enter into myriad of non-EU/EEA agreements. Consulting the EU Treaties Office Database reveals bilateral and multilateral treaties with Canada, US, Brazil, China, and many other countries and regions around the globe. Among shared interests are agriculture, commercial policy, competition, culture, economic and monetary affairs, education, energy, environment, fisheries, food safety, foreign and security policy, information society, research and innovation, taxation, trade and transport (see also European Commission 2009).

These treaties represent economic, social and market opportunities that contribute to the normalization of mobilities as the way forward. On the face of it, these developments appear as "a promising sign of global newness and future-making" (Gill et al. 2011; Adey 2010), a form of cosmopolitanism, manifested in enhanced freedoms and opportunities to prosper across the globe in an era of declining nation states. However, we are warned against associating mobility uncritically with liberation from nationhood and widespread personal freedoms (e.g. Cresswell 2006). Borders and other checkpoints continue to delineate and connect, divide and categorize persons and cargo, reject, admit, contain and alienate, and they are contentious in political debate.

### *6.2.1 Politics of Cohesion*

Economic, social and territorial cohesion in Europe is seen as essential for the stability of the Single Market, and the European Union (EU) has a pivotal role to play in staging its moral character on the basis of the EU constitution. The Union shall "offer its citizens an area of freedom, security and justice without internal frontiers, [...] combat social exclusion and discrimination, and [...] promote social justice and protection" as well as solidarity (European Union 2010, Art.3 of Common Provisions p. 17). But, social and territorial cohesion is subject to doubt when the distribution of opportunities, wealth and welfare is unequal among member states, of which some are grappling with internal unrest. States in Europe are not equal with respect to their governance traditions and development. There are significant differences to be observed between the northern, central and southern territories. Some states are richer than others and some more influential in cooperation on matters of economy, markets, innovation and societal affairs.

These differences influence the ways in which the mobility of persons is looked upon and managed. The current amount of mass irregular travel and nomadic labour was not anticipated in the early days of the Single Market and the entitlement to mobility *for all* has not always been welcome, not even within Schengen. The second paragraph in Article 2 of the Schengen agreement allows for unilateral reintroduction of border checks, i.e., spot checks, which are periodically invoked in relation to events or *exceptional circumstance* seen as a threat to security (Council of the European Communities 2000; Apap and Carrera 2003). For example, migration anxieties surfaced in 2011, in relation to the so-called Arab spring (Scuzzarello and Kinnvall 2013) and Italians saw an influx of migrants from the conflict zones on their shores. But, a decision to issue temporary residence permits to Tunisian refugees triggered patrol on the French-Italian borders to stop that migration going further north. Anti-immigrant sentiments were also heard in Denmark at the time, however, with focus not only on migrants from the Maghreb region, but from Poland and the Baltic states. Similar anxieties were evident in warnings of disproportionate migration resulting from the enlargement of the EU to eastern regions—concerns, not entirely void of scepticism about social-cultural kinship (Kvist 2004; Bauer and Zimmermann 1999), and the delay in opening the doors to citizens of the new member states allegedly put into effect to protect social infrastructures.

Scuzzarello and Kinnvall argue that these kinds of events show how quickly borders are reclaimed and ideas of a collective self and national identity reinstated in political and media discourse. This happens periodically within the EEA and Schengen and one can argue that Europe remains geographically, culturally and economically divided regardless of the four freedoms lubricating the Single Market and relations beyond.[4] Those from outside the EEA and the Americas, in particular, from where asylum-seekers and economic migrants originate, are typically neither

[4] Crossing patrolled borders between Norway and Denmark soon after the conflict with the Islamic State began, also illustrates some of the complications faced within Schengen when EEA citizens get involved in conflicts.

wanted nor welcome either. They are not the *ideal foreigners*, the educated and mostly white who arrive with social, cultural and economic capital to travel, study, work, consume and even settle (see Tyler 2006, on the UK case). It is too easy to idealize mobility. There are opportunities *for some* to study and accept job offers across state lines and there is a tourist trade aimed at those with disposable incomes. Frequent long-distance travel remains privileged however, and so is effectively the so-called *global workforce*. As Castells puts it,

> [w]hile capital flows freely in the electronic circuits of global financial networks, labor is still highly constrained, and will be for the foreseeable future, by institutions, culture, borders, police, and xenophobia (Castells 2010: 247).

If these observations are anything to go by, only the privileged labour force of highly specialized professions, are living their lives across national borders with relative ease and comfort (also Cooper and Rumford 2013). At the same time, an intensification of labour mobility and tourism was seen in all regions of the world toward the end of the last century (e.g. Massey et al. 1998) and the trend continues. Being *open for business* opens the doors to both legitimate and illegitimate flows. Mobile capital benefits from an authority structure that defends its mobility and privilege, but radicalized hostility toward market democracies and corporate enterprise calls for new priorities in thinking about the cohesion of this and other markets. The securitization rationale has adopted narratives that depict particular kinds of flows as trusted against flows that are a threat to the socio-economic order—narratives that mediate belonging, set the threshold for inclusion, and demand that mobility data is made available to authorities in the name of security.

## 6.3 Mobility Controls and the *Fifth Freedom*

Drawing on mobilities research, one can argue that agreements like the EEA and Schengen, constrain the freedoms of movement as much as they facilitate them but, also, the movement of data deserves closer attention. Modern states are increasingly more mobile to keep up with persons, capital and commodities (Beck 2005, 2008). They need to keep up with an economy that is *informational* as well as globally organized (Castells 2010), drawing together materials, commodities, markets and labour, along with the business models and the organizational management for which the free movement of data is an absolute necessity. Within the EEA, the free movement of data is effectively the *fifth freedom*, and even if not explicitly recognised as such, an *Informational EEA* is obliged to protect that freedom. This is evident, for instance, in the preamble of the GDPR, stating the aim to protect the free movement of data across the internal market.[5]

---

[5]The title of the GDPR also states that it is a regulation "on the protection of natural persons with regard to the processing of personal data and on the free movement of such data" (http://eur-lex.europa.eu/legal-content/EN/TXT/PDF/?uri=CELEX:32016R0679&from=EN).

Historically, all sorts of data are processed for supervision and operational surveillance. For example, the types of surveillance on persons manifested in public registries have heritage in record keeping and investigative techniques for social security entitlements, tax collection, election management, healthcare, border control, intelligence and ordinary policing. Outsourcing the processing of these data to private and corporate agencies has gone on for years, but the more advanced surveillance systems – including biometry – have a history as well in banking, industry, insurance and commerce (Lodge 2006).

A significant amount of data on persons is now in circulation throughout the whole economy (and growing), either identifiable or clustered by type in reference to locality, mobility, biophysical profiles, action, preference, and behaviour.[6] Huge amounts of data on persons are processed by the service industries and in marketing with the bulk of it now generated by individuals themselves (Hoback 2013). Targeted marketing and so-called personalized services rely on consumers volunteering data and, as the argument goes, blanket restrictions will harm competitiveness by undercutting existing business models, back-office efficiency, new service discovery and the delivery of customer service quality (European Parliament 2012).

In a world where mobilities are imperative if markets are to succeed, ICTs are essential to market productivity as well as in the surveillance of marketing efficiency. Research into targeted marketing has led the way for decades on how and why mobilities should be traced and monitored (e.g. Phillips and Curry 2003). It has shown the difficulty in relating geographically localized populations (geodemographics) to sociological ideals about identity, lifestyle choice, circumstance or preference. These failings in defining localized groups suggested that knowledge of persons would be better established by following them around. Similar advantages are surfacing more recently with the growing use of wearable sensors, i.e., to establish knowledge on biophysical and health-related profiles. *Checkpoints* are monitored to target any such information with reasonable accuracy, all over the internet, the use of mobile and smart phones, sensors, service systems of sorts, machine-readable cards or any other transaction-type or registered event (Bigo 2007).

The analogy here with border control describes how any place, physical or virtual, can be turned into a *checkpoint*. Individuals are intercepted, asked to identify themselves and clarify their travels, activities and transactions, aspirations, intentions, the goods and keepsakes that travel in their name. These practices exemplify both *facilitation* and *restriction* of mobility, however, private and corporate enterprise manage all manner of such checkpoints at their own discretion and the idea that people's intentions can likewise be detected and intercepted has gained significant momentum. Commercial interests and economic gains of the software and

[6] It can be argued that data clustering by types anonymizes data, i.e., the data controller is not interested any one person's official identity, only that such and such a person is of such and such a *type*. This is debatable however, since the goal is typically to be able to intercept *types of persons* via email, web browsing, phone messaging or old fashioned mail, all of which is ultimately traceable back to actual individuals.

communications industries give rise to suspicion here that governments are not entirely in control of how to regulate these developments and the *fifth freedom* has neither been conspicuous nor transparent to publics at large. Publics are still not critically involved in assessment and decision-making on matters concerning the collection and processing of personal and potentially sensitive data, nor the extent to which surveillance can be practized by private and corporate agencies for commercial purposes and economic gain.

### 6.3.1 Self-Censoring Publics

The securitization agenda has shifted the focus in law enforcement, security and related operations, toward pervasive surveillance for investigative purposes which takes advantage of the data flows in the service industries. Surveillance is now "a feature of everyday life, at work, at home, at play, on the move" (Lyon 2003: 13), and the Snowden revelations provide stunning insights into precisely the extent to which surveillance takes place and how publics are targeted (Poitras 2014). The latest technological advancements are mobilized to that effect with the shift toward preventative scenarios involving research into the remote monitoring of persons in large numbers. Smart cameras are attached to video content analysis, combined with biometric techniques for facial recognition, behavioural pattern recognition and remote sensing of physiological states. It is unclear however, what is actually achieved with all this data mining, statistical categorizations of *dangerous* groups, profiling of individuals, and ICT-assisted predictions of future events (on profiling see Hildebrandt and Gutwirth 2008; Hildebrandt 2008).

Serious doubts have been raised about the investigative and preventative scenarios and it is yet unclear whether or not such systems eventually come online to achieve what designers have hoped for (e.g. the European project ADABTS (2013)[7] and the US Future Attribute Screening Technology (FAST) – Interactive and Passive Programs[8]; also Hornung et al. 2010; Schlanger (reviewing official) 2011). The technology is not entirely reliable and extensive ethical and legal reviews have called for adjustments, with some projects abandoned entirely or scaled down to small location-specific applications. For example, an impact assessment for FAST from 2011 states that,

> FAST is still in the research phase; it is *not currently, and may never become, operational.* Nonetheless, it seems appropriate, as the Department works on this research, to ensure that deployment, *if it ever occurs*, would not diminish the public's civil rights or civil liberties. See 6 U.S.C. 111(b)(1)(G). (Schlanger 2011: 2, [*emphasis added*]).

Apparently, even those who work within the agencies investing significant sums in this research are not clear whether or not the envisioned systems are realizable. As

[7] See https://www.informationsystems.foi.se/main.php/ADABTS_Final_Demo.pdf?fileitem=7340169

[8] See http://en.wikipedia.org/wiki/Future_Attribute_Screening_Technology

general investigative tools however, advanced surveillance systems along with the data logging and profiling that underpins existing business models, management and discovery in the service industries, are providing unprecedented access to data across state lines for search-and-match purposes (Hoback 2013).

The danger here is losing sight of opportunities to meaningfully object to the pursuit of executive powers to intercept data flows and logs in these systems. As Snowden has put it (Poitras 2014), publics are self-censoring if they live with the expectation of being watched all the time. Indeed, it appears that speaking of liberties and freedom has been captured in the notion of *privacy* in the very particular sense that *not having privacy* is a loss of a person's agency because they no longer feel free to express what they think (Applebaum, cf. Poitras 2014). Privacy is here primarily linked to freedom of speech and expression.[9]

> If we don't have our right to privacy, how do we have a free and open discussion? What good is the right to free speech if it isn't protected in the sense that you can't have a private discussion with somebody else about something you disagree with. Think about the chilling effect that that has. (Levison at the hearings in EU parliament on NSA surveillance of EU citizens and companies, Sept. 2013, cf. Poitras 2014, 1:38:45–1:39:05).

It may seem that a distinction between *private* and *public* needs to be ascertained here, and how the two intersect. Any given communication can be more or less private or public, and not obvious when indeed there is freedom to express one's opinion given what the incentives are to self-censor, for example, of not being associated with certain persons, politics and events, and not being put on some watch list or other. The same can be said about avoiding physical attire and behaviour that may give rise to alarm. But, distinctions of public and private are not at issue here, rather, the conditions by which self-censoring becomes a feature of everyday life. As it stands, individuals and groups are posed against a coalition of agencies, whose personnel are learning how the most advanced technologies can assist them in detection—how security and surveillance systems *aimed at everyone* can assist in establishing ever more detailed knowledge on *who they are* and in preventing certain kinds of events from happening. This positioning may well sit on a continuum with older investigative techniques used by state agencies and constabularies to put on a display of power, to uphold the rule of law and investigate acts of crime and terror. There are well known pros and cons in the application of ICTs and biometry as part of intelligence gathering, and such practices are continuously negotiated in reference to the rights and liberties of persons, what is proportionate, justified, and so on.[10] What is at issue here is rather the tensions that arise when such a versatile materiality of economic competitiveness and growth, *data,* is exposed to *excep-*

[9] Article 19 of the Universal Declaration of Human Rights (The United Nations 1948) states that everyone has the right to freedom of opinion and expression. However, widespread self-censorship is a well-known phenomenon, notably in recent US history during the McCarthy era and in recent European history east of the Iron Curtain.

[10] An issue repeatedly foregrounded during the ICTethics expert workshop on "Human Security in the context of Ambient Intelligence", Nov 2010, Leeds, UK (http://neicts.lancs.ac.uk/workshops.htm#workshop3).

*tional circumstances*, whereby personal data – in fact any data – can be subpoenaed or otherwise intercepted for purposes no one can truly challenge. The GDPR is an attempt to improve upon the protection of personal data, of clarifying the ramifications of data processing and retention in an era of increasingly more powerful technologies.[11] It has no force however, in challenging data operations relating to criminal investigations, security or terrorist prevention programmes—areas of activity where publics need protection the most if indeed the practical utility of securitization is primarily to target political resistance couched as *threat to public safety and the social order*. Publics may have no choice left but to self-censor while there are still good reasons to speak up and act in defiance of authorities on a range of deeply contentious issues, including social injustices, decisions on warfare, the direction of the socio-economic system and the purposes of scientific, technological and other societal innovations.

## 6.4 Keeping *What* Safe?[12]

The four freedoms of movement, as stated in the EEA agreement, do not immediately signify how mobilities leave behind traces of transactions that open the doors to pervasive surveillance. It is not obvious either what the fifth freedom signifies in terms of how to protect individuals with regard to the processing of personal data. This chapter began by claiming that the securitization agenda emerges in consequence of *insecurities*, albeit, there is persistent lack of clarity on why securitization – as it proceeds – seems inevitable. On the face of it, securitization is justified in reference to enemies descending on first-world democracies and hiding *among* the citizenry. Indeed, the agenda has been particularly instrumental in cultivating migration anxieties, scepticism over entitlement to mobility and in framing the problem of *threat* more generally in terms of an imperative to detect and identify those who are dangerous.

There is nothing to indicate other than a continuity of familiar old-age problems of oppression and resistance, corruption, exploitation, abuse, crime and terror, only that these problems appear in new guises and involve the latest technologies appropriated by enemies and allies alike. Narratives of threat and emergency mask this, along with a whole host of considerations that have a history of being ideological no-go zones. For example, openly debating with publics the very purpose and direction of first-world leadership in economic affairs, innovation and market competi-

[11] As this is written, the GDPR has just entered into force after years of preparations and thousands of amendments, of which many were intended to better safeguard business models, economic and innovation competitiveness. Provisions will apply in all EU member states from 25 May 2018.

[12] Some of these concluding remarks I owe to an inspiring meeting (Jan 2014) with the Icelandic parliamentarian, Birgitta Jónsdóttir, and internet pioneer, Guðmundur Ragnar Guðmundsson, in particular, their thoughts on digital self-defence in contemporary consumer culture, and the lack of teeth in the GDPR with respect to its jurisdiction and scope.

tiveness, might result in 'unpalatable' demands to do things differently. Such debates are simply not had in organized ways and reported on in mainstream media, involving key decision-makers along with citizen juries or other recognised methods of public engagement.[13] Openly debating whether or not *our leadership* is sustainable, in particular, if it erodes the cohesion it is meant to aim for, may call for an honest reflection on the rhetoric of *leadership* and *keeping us safe*. When used by leaders and mainstream media, such rhetoric draws the attention away from state-sponsored aggression in defending vested interests in goods, services, capital, mobile labour *and data* (also Herman and Chomsky 1988 on manufacturing consent). Furthermore, narratives of threat and emergency call for immediate intervention. Under the circumstances, government executives manipulate the legislature to rush through emergency laws, bypass the judiciary or ignore both, while going ahead with programmes that are shrouded in secrecy (Harding 2014; Poitras 2014).[14]

It is at this juncture that one can see that the *insecurities* in question here, centre primarily on *publics-as-threat*, against which the securitization agenda has acquired its *practical utility*. It is also at this juncture that the pivotal role of ICT-based innovation finds its relevance, as previously stated in this chapter, in conjunction with long-standing practical uses of surveillance and security technologies in industry, banking, trade, commerce and other service sectors. The intensity of contemporary surveillance has been much more political than fending off nebulous enemies who can *harm us* at any moment. Intelligence, observation and quantification techniques are used to code and classify persons according to definitions that are interpretatively flexible, hence, these techniques are ideal to identify any *publics-as-threat*, for example, those who mobilize against uncontested innovation in industries and markets for growth and competitiveness (Welsh and Wynne 2013).

Even after the Snowden revelations began, little is done in the way of communicating and critically debating in the public domain the purposes with which intelligence and security operations proceed and, for Europe in particular, the extent to which the GDPR actually protects persons. Given the strength of the official rhetoric of threat and emergency, and the silencing of opinion through self-censorship, this is not surprising. But, democratically elected bodies should be held to account for the ways in which ICT-driven surveillance operations are allowed to propagate as the inevitable way forward in defending *our way of life*. There is no evidence that advancing ICTs alone will help remedy dissent and hostility and cultivate cohesion.

[13] Debates that really shake the core assumptions on which the economy rests, first-world leadership, innovation, security, etc., are held in small ways, involving whistleblowers, critics and disillusioned members of established institutions who find voice on minor media outlets, typically referred to as left wing or 'too' radical to be on par with mainstream communication, e.g., Democracynow! (http://democracynow.org), Counterpunch.(http://www.counterpunch.org/) and New Statesman (http://www.newstatesman.com/).

[14] There are many attempts over the past decade to back-track on executive powers, e.g., in summer 2015, a disquiet in the UK parliament over the lack of transparency regarding the reach of executive powers, including the GCHQ, without the checks and balances of a three-tiered government. The US Patriot Acts have also been revisited, the US surveillance law, the legality of the Guantánamo Bay detention camp, and much more.

Rather, there is good evidence that the software and communications industries have significant investment in the securitization agenda, while the question remains to what extent the service industries and the markets should take part, given that their investments (e.g., targeted marketing) depend on data-driven practice.[15] Authorities can very effectively leave most of the monitoring and measuring of everyday mobilities to private and corporate enterprise and issue subpoena when needed.[16] Service providers can refrain from retaining data (metadata or content data), especially identifiable data, but authorities can still intercept all transactions under circumstances removed from democratic accountability. The only alternative then is to close up shop (Levison, cf. Poitras 2014, 1:37.02-1:38:20)

Persistent lack of clarity on the purposes of mass-surveillance, *what exactly is kept safe* and who is actually in charge, points to a need for critical reflection and debate, albeit, public debates are not likely to lead the way forward in a climate of self-censorship and, to paraphrase Snowden, when the boundaries of intellectual exploration have been limited (26:54–26:58).[17] One can also ask why publics at large are not more encouraged to consider the pre-eminence of their informational selves and digital citizenship. For example, we have yet to see widespread education programmes from primary school level and upward, teaching on these topics. As it stands, the majority of publics indulge the complacency which is part and parcel of free market democracy—the comforts of consumption, of ready-to-hand products and services, leaving us all vulnerable to marketeers and government agencies pushing uncontested agendas. The majority of publics may not see a risk to their mobility or to their "private and family life, home and communications" (European Communities 2007: Art. 7), and publics are not commonly educated in electronic and digital self defence. There are no provisions for such defence in the European or US constitutions,[18] but guarantees of interoperability between industry software and citizens' initiatives to build digital and informational safety around themselves are still a legitimate demand. As it stands, the software and communications industries cannot be obliged to use open-source encoding of security software, say, of encryption algorithms, which means there are no adequate tools to keep industry

[15] In the aftermath of the Snowden revelations, the big industry players in ICTs are publishing on this, e.g., IBM on regulatory affairs and Microsoft on public policy agendas, and there has been a gradual shift in recent years toward seeing privacy sensitivity as a resource for profitable innovation, e.g., in selling to prospective customers privacy-by-design and protection-by-design products and services.

[16] Arrangements like that between government agencies and private enterprise are indicated in the Snowden revelations. Yahoo, Google, Facebook, YouTube and Skype were implicated in the PRISM programme (NSA) and the UK GCHQ accessing data on persons through that programme, thereby circumventing legal procedures.

[17] See also interview with Glenn Greenwald and Laura Poitras (9 June 2013), in which Snowden elaborates one of his key arguments that publics need to be consulted and somehow involved in decisions on the extent to which personal data handling is publicly acceptable, in particular, in light of the self-censoring going on. Available at http://www.theguardian.com/world/video/2013/jun/09/nsa-whistleblower-edward-snowden-interview-video (visited Aug 2016).

[18] It is of some curiosity that US citizens have a constitutional right to armed defence of their person and property but no such right to defence of their informational person and property.

software in check, in particular, for vulnerabilities to snooping and other illiberal practices.

What is publicly acceptable should be put to the test here against the practical utility of the securitization agenda, taking into account that public acceptability risks compromise when ill-defined threats are instrumental in encouraging people to accept personal data processing without asking what it is for (see European Commission 2008). The *fifth freedom* may certainly warrant protection, however, as long as there are serious blunders in the governance of executive powers, the very idea of *having control* of personal data protection by way of law-making (or of one's privacy for that matter), is largely void of meaning. Having educated debates across the Atlantic on the importance of privacy, liberty and self-determination in the making of first-world democracies, on striking the right balance, and so on, is equally futile when any data that can be collected is, in all likelihood, collected eventually by some agency, overtly or covertly, processed, disseminated and retained for purposes that cannot be contested. These topics have yet to enter into dialogue that can exercise legitimate powers of participation and persuasion—to engender a critical re-framing and re-evaluation of the securitization agenda with a view to much greater checks on executive powers along with open debate and deliberation on the kinds of future societies first-world citizens can imagine for themselves in relation to the rest of the world.

## References


Adey, Peter. 2010. *Mobility*. London: Routledge.

Agamben, G. 2005. *State of exception*. Chicago: University of Chicago Press.

Amoore, Louise. 2006. Biometric borders: Governing mobilities in the war on terror. *Political Geography* 25(3): 336–351.

Apap, Joanna, and Sergio Carrera. 2003. Maintaining security within borders: Towards a permanent state of emergency in the EU? *CEPS Policy Briefs 41, October 2003*. [Policy Paper].

Bauer, Thomas, and Klaus F. Zimmermann. 1999. *Assessment of possible migration pressure and its labour market impact following EU enlargement to central and Eastern Europe – A Study for the Department for Education and Employment, United Kingdom*. IZA Research Report No. 3. http://www.iza.org/en/webcontent/publications/reports/report_pdfs/report_pdfs/iza_report_03.pdf. Accessed Aug 2016.

Beck, Ulrich. 2005. *Power in the global age: A new global political economy*. London: Polity.

Beck, Ulrich. 2008. Mobility and the cosmopolitan perspective. In *Tracing mobilities: Towards a cosmopolitan perspective*, ed. Weert Canzler, Vincent Kaufmann, and Sven Kesselring, 25–36. London: Ashgate.

Bigo, Didier. 2007. Mobility controls and new technologies. In *Are you who you say you are? The EU and biometric borders*, ed. Lodge Juliet, 9–14. Nijmegen: Wolf Legal Publishers (WLP).

Carothers, Thomas. 2003. Promoting democracy and fighting terror. *Foreign Affairs*, January/February. http://www.foreignaffairs.com/articles/58621/thomas-carothers/promoting-democracy-and-fighting-terror. Accessed Aug 2016.

Castells, Manuel. 2010. *The rise of the network society: The information age, economy, society and culture, volume 1 (2nd edition with a new preface)*. Wiley-Blackwell.

Cooper, Anthony, and Chris Rumford. 2013. Monumentalising the border: Bordering through connectivity. *Mobilities* 8(1): 107–124.

Council of the European Communities. 2000. The Schengen acquis. *Official Journal* L 239, 22.9.2000. http://eur-lex.europa.eu/legal-content/EN/TXT/HTML/?uri=OJ:L:2000:239:FULL&from=EN. Accessed Aug 2016.

Cresswell, Timothy. 2006. *On the move: Mobility in the modern Western world*. Taylor & Francis.

Dalacoura, Katerina. 2005. US democracy promotion in the Arab Middle East since 11 September 2001: A critique. *International Affairs* 81(5): 963–979.

European Commission. 2008. Data protection in the European Union – Citizen's perceptions (Flash Eurobarometer 225), Feb. 2008. http://ec.europa.eu/public_opinion/flash/fl_225_en.pdf. Accessed Aug 2016.

EU External Action Service. 2011. *Association Agreements*. European Community/European Union. http://eeas.europa.eu/association/docs/agreements_en.pdf. Accessed Aug 2016.

European Communities. 2007. Charter of fundamental rights of the European Union. *Official Journal* C 303/01. http://eur-lex.europa.eu/legal-content/en/TXT/?uri=CELEX%3A12012P%2FTXT. Accessed Aug 2016.

European Parliament. 2012. *Data protection review: Impact on EU innovation and competitiveness (study)*. Directorate General for Internal Policy. Policy Department A: Economic and Scientific Policy. PE 492.463. http://www.europarl.europa.eu/RegData/etudes/etudes/join/2012/492463/IPOL-ITRE_ET(2012)492463_EN.pdf.

European Parliament and the Council of the European Union. 2016. Regulation (EU) 2016/679 of the European Parliament and of the Council of 27 April 2016 on the protection of natural persons with regard to the processing of personal data and on the free movement of such data, and repealing Directive 95/46/EC (General Data Protection Regulation). *Official Journal* L 119, 4.5.2015. http://eur-lex.europa.eu/legal-content/EN/TXT/PDF/?uri=CELEX:32016R0679&from=EN. Accessed Aug 2016.

European Parliament Fact Sheets. 2000. 6.3.2. *The European Economic Area (EEA)*. European Parliament. http://www.europarl.europa.eu/factsheets/6_3_2_en.htm. Accessed Aug 2016.

European Union. 2010. Consolidated versions of the Treaty on European Union and the Treaty on the Functioning of the European Union & Charter of Fundamental Rights of the European Union. *Official Journal* C 83, 30.3.2010. http://eur-lex.europa.eu/JOHtml.do?uri=OJ:C:2010:083:SOM:EN:HTML. Accessed Aug 2016.

Gill, Nick, Javier Caletríob, and Victoria Mason. 2011. Introduction: Mobilities and forced migration. *Mobilities* 6(3): 301–316.

Harding, Luke. 2014. *Snowden files*. London: Guardian Books.

Herman, Edward S., and Noam Chomsky. 1988. *Manufacturing consent: The political economy of the mass media*. New York: Pantheon Books.

Hildebrandt, Mireille. 2008. Profiling and the rule of law. *Identity in the Information Society* 1: 55–70.

Hildebrandt, Mireille, and Serge Gutwirth (eds.). 2008. *Profiling the European citizen: Cross-disciplinary perspectives*. Dordrecht: Springer.

Hoback, Cullen (Director). 2013. *Terms and conditions may apply*. USA: Hyrax Films.

Hornung, Gerrit, Monika Desoi, and Matthias Pocs. 2010. Biometric systems in future preventive scenarios – Legal issues and challenges. In *Proceedings of the special interest group on biometrics and electronic signatures (BIOSIG 2010) Darmstadt, Germany 9–10 September 2010*, ed. A. Brömme, and C. Busch, 83–95. Gesellschaft für Informatik e.V. (GI).

Jordan, Bill, and Philip Brown. 2007. Migration and work in the United Kingdom: Mobility and the social order. *Mobilities* 2(2): 255–276.

Kvist, Jon. 2004. Does EU enlargement start a race to the bottom? Strategic interaction among EU member states in social policy. *Journal of European Social Policy* 14(3): 301–318.

Lodge, Juliet. 2006. *Communicating (in)security: A failure of public diplomacy*. CHALLENGE: Research Paper no.3. Centre for European Policy Studies, Brussels. http://aei.pitt.edu/7403/2/7403.pdf. Accessed Aug 2016.

Lyon, David (ed.). 2003. *Surveillance as social sorting: Privacy, risk, and digital discrimination*. London/New York: Routledge.

Massey, Douglas, Joaquín Arango, Graeme Hugo, Ali Kouaouci, Adela Pellegrino, and J. Edward Taylor. 1998. *Worlds in motion: International migration at the end of the millennium*. New York: Oxford University Press.

Meltzer, J.P. 2014. *The importance of the internet and transatlantic data flows for U.S. and EU trade and investment (working paper 79)*. The Brookings Institution: Global Economy & Development Program.

Obama, Barak H. 2011. President Obama addresses the British Parliament, Westminster Hall, 25.5.2011 London, United Kingdom. http://news.bbc.co.uk/democracylive/hi/house_of_commons/newsid_9495000/9495513.stm . Accessed Aug 2016.

Phillips, D., and M. Curry. 2003. Geodemographics and the changing spatiality of the local practice. In *Surveillance as social sorting: Privacy, risk, and digital discrimination*, ed. D. Lyon, 137–52. London: Routledge.

Poitras, Laura (Director). 2014. *Citizenfour (documentary)*. Praxis Films in association with Participant Media and HBO Documentary Films; in co-production with Bertha Foundation, Britdoc Circle, Channel 4, NDR and BR media.

Schlanger, Margo, Reviewing Official. 2011. Civil Rights/Civil Liberties Impact Assessment. Future Attribute Screening Technology (FAST) – Interactive and Passive Programs. USA Department of Homeland Security. http://www.dhs.gov/xlibrary/assets/crcl/crcl-assessment-fast.pdf. Accessed Aug 2016.

Scuzzarello, Sarah, and Catarina Kinnvall. 2013. Rebordering France and Denmark: Narratives and practices of border-construction in two European countries. *Mobilities* 8(1): 90–106.

Sheller, Mimi, and John Urry. 2006. The new mobilities paradigm. *Environment and Planning A* 38(2): 207–226.

Shove, Elizabeth. 2002. *Rushing around: Coordination, mobility and inequality*. Draft paper for the Mobile Network meeting, Oct. 2002, Department for Transport, London. http://www.lancaster.ac.uk/staff/shove/choreography/rushingaround.pdf. Accessed Aug 2016.

Stephenson, Marcus L. 2006. Travel and the 'freedom of movement': Racialised encounters and experiences amongst ethnic minority tourists in the EU. *Mobilities* 1(2): 285–306.

The United Nations. 1948. The Universal declaration of human rights, UN Publications. http://www.un.org/en/universal-declaration-human-rights/index.html. Accessed Aug 2016.

Tyler, Imogen. 2006. 'Welcome to Britain': The cultural politics of asylum. *European Journal of Cultural Studies* 9(2): 185–202.

Urry, John. 2002. Mobility and proximity. *Sociology* 36(2): 255–274.

Urry, John. 2007. *Mobilities*. London: Polity Press.

Welsh, Ian, and B. Brian Wynne. 2013. Science, scientism and imaginaries of publics in the UK: Passive objects, incipient threats. *Science as Culture* 22(4): 540–566.

# Chapter 7
# How Do Violations of Privacy and Moral Autonomy Threaten the Basis of Our Democracy?

**Katrin Laas-Mikko and Margit Sutrop**

**Abstract** Behavior detection technologies are currently being developed to monitor and manage malintents and abnormal behavior from a distance in order to prevent terrorism and criminal attacks. We will show that serious ethical concerns are raised by capturing biometric features without informing people about the processing of their personal data. Our study of a range of European projects of second-generation biometrics, particularly of *Intelligent information system supporting observation, searching and detection for security of citizens in urban environments* (INDECT) and *Automatic Detection of Abnormal Behaviour and Threats in Crowded Spaces* (ADABTS), shows that violations of privacy put several other values in jeopardy. We will argue that since privacy is in functional relationship with other values such as autonomy, liberty, equal treatment and trust, one should take this into account when limiting privacy for protecting our security. If indeed it should become necessary to restrict our privacy in specific situations, thoughtful consideration must be given to other ways of securing the values that form the foundation of our liberal democratic society.

**Keywords** Privacy • Security • Autonomy • Democracy • Second-generation Biometrics • Behavior detection • Surveillance technology • Dataveillance • Function creep • INDECT • ADABTS

---

This paper is a slightly revised version of the article published in *Trames: Journal of the Humanities and Social Sciences* vol. 16, no. 4, 2012. It has been published with the permission of the Estonian Academy Publishers.

K. Laas-Mikko (✉) • M. Sutrop
Institute of Philosophy and Semiotics, University of Tartu, Tartu, Estonia
e-mail: Katrin.Laas@ut.ee; Margit.Sutrop@ut.ee



A. Delgado (ed.), *Technoscience and Citizenship: Ethics and Governance in the Digital Society*, The International Library of Ethics, Law and Technology 17,
DOI 10.1007/978-3-319-32414-2_7

## 7.1 Introduction

Privacy is an important value, but protecting it continues to become more difficult. One reason is that another value, security, seems to be increasingly in danger, and its defense appears to demand significant curbing of privacy. The high rate of population growth, significantly increased mobility, and the development of information and communication technologies entail the elevation of security to the position of supreme value, in the name of which people are willing to limit privacy or relinquish it altogether.

In the context of such noble goals as creating a safe world, the protection of privacy may well turn out to be illusory. There are numerous predictions that the "Fight against terror 'spells the end of privacy'" (Travis 2009). Several recent trends and government initiatives which aim to enhance the capability for surveillance and detecting potential criminals or terrorists also feed growing apprehensions that the future may not bode well for the right to privacy.[1]

The problem is that what is being endangered is not privacy alone; since privacy supports a range of other values, limitations on privacy can also place these other values at risk. As pointed out by several authors (Gavison 1980; Kupfer 1987; Solove 2007), privacy promotes liberty, autonomy, selfhood, and human relations, and furthers the existence of a free society. Therefore, in a democratic state one should continually be posing the question, what is the price of prioritizing security? The main purpose of our paper is to urge upon us the need to weigh carefully whether we are actually willing to relinquish privacy and a host of other values in the name of security. Of course, security is a crucial matter, but the means we use to ensure it should be proportional to the greatness of the potential threat. We should also consider whether those values that have previously been maintained by privacy can be protected in some other way.

Indeed, it is paradoxical that decisions are being made to limit privacy in order to protect a democratic society and guarantee its security, while these same means of protection can erode that society (more effectively than attackers might have done) by undermining its basic values. This reminds us of the satirical film, "Team America: World Police", where in the name of capturing a few terrorists, the Louvre and the Eiffel Tower are blown up, along with other landmarks essential to our culture.[2]

In what follows we will first analyze the concept of privacy, beginning with the functions it fulfils in liberal democratic societies. Subsequently, we will examine some specific examples of second-generation biometric projects which raise serious ethical concerns: behavior detection without informing people about the processing

---

[1] For example see the following with regard to cell phone surveillance in the US. http://www.nytimes.com/2012/07/15/opinion/sunday/the-end-of-privacy.html, security laser scanners used in airport controls http://washington.cbslocal.com/2012/07/11/new-homeland-security-laser-scanner-reads-people-at-molecular-level/ etc.

[2] "Team America: World Police" Details 2004, USA, Cert 15, 98 mins, Animation/Satire, Dir. Trey Parker.

of their personal data violates not only their privacy, but also their moral autonomy and other values. The aim of this paper is to show that we should not give up our privacy too easily; after all, privacy is a necessary condition of autonomy and democracy, without which our society will lose its foundation.

## 7.2 The Value of Privacy

Privacy can be described as limited to the 'sphere' surrounding the person, within which that person has the right to control access to himself or herself. Consequently, privacy is applicable only within certain boundaries (not necessarily or only spatial) that surround that person (Persson and Hansson 2003, p. 61). We support a normative concept of privacy, where privacy means the person's right to decide to what extent other persons can access and use information concerning him or her, and who those persons are who have access to his or her physical body; those who access and use physical/intimate space surrounding the person.

Privacy can be an intrinsic value, while also fulfilling many important goals. In her article "Privacy and the Limits of Law" (1980) Ruth Gavison has shown that when speaking of the functional/instrumental meaning of privacy, we begin with the assumption that privacy is concerned with developing or preserving something that is desired. Gavison distinguishes between functions that privacy has with respect to the individual and those that concern society as a whole. In order to determine what functions privacy has with respect to the individual, we must consider what we deem important about being a person. Many Western theorists (Gavison 1980; Kupfer 1987; Häyry and Takala 2001; Rössler 2005) have indicated that the primary task of privacy with respect to the individual is to protect his or her autonomy. For example, Joseph Kupfer (1987, pp. 81–82) claims that privacy is a necessary (though not the only) condition for the development of an autonomous 'I' or self. One of the most thorough treatments of privacy has been presented by Beate Rössler in her book, "The Value of Privacy" (2005), which centers on the question of why we value privacy. Rössler endeavors to show that "privacy in liberal societies is valued and needed for the sake of individual liberty and autonomy, that is, for the sake of both freedom for each individual to fulfill himself, and thus ultimately for the sake of a life that is rewarding" (2005, p. 44). She describes privacy as the ability to control 'access' in the physical or metaphorical sense to one's personhood, enabling the practice of autonomy; decisional privacy, meaning free behaviour and action with respect to matters pertaining to oneself; informational privacy, which means controlling what other people can know about oneself; and local privacy, designating the protection of one's own space for purposes of self-evaluation and intimate relationships. In the context of the development and application of new technologies, our main concern is with informational privacy, a person's control over the access and use of information about himself or herself.[3]

[3] See Adam Moore's (2008) similar definition of informational privacy.

The answer to the question, what kind of society we desire depends on what we believe to be important to live a good life. Society should create opportunities for the flourishing of the individual. Preconditions for the self-realization of the autonomous individual are a liberal democracy and a pluralistic society, where everyone can live according to his or her chosen model of the good life. The reverse is also true: a democratic society presupposes an autonomous individual who can determine for themselves what constitutes the good and to choose the means of achieving this.

However, one should not only be able to determine what is good for oneself, but also the common good we all share. Collective benefits – security, peace, order, justice – pertain to all members of society and their existence depends on reciprocity. Since there are all kinds of obstacles standing in the way of achieving these conditions, such as subjective interests and continuous competition over scarce externally obtainable goods, the question is, how should the common good be defined. There are also differences in the extent of individuals' abilities to participate in deliberations concerning the meaning of the common good.

Our suggestion is to learn from Aristotle's political philosophy, in which a clear, cogent, and persuasive effort is made to bind together the common good and autonomy. The ground of Aristotle's political philosophy is his belief that "the state is by nature clearly prior to the family and to the individual, since the whole is of necessity prior to the part" (Aristotle 1996, p. 1253a19) but none of these terms stands independently of the other. The good of the republic is imbricated with the good of each citizen and the citizens experience the goodness of their republic in their everyday lives. For Aristotle the common good refers to "a good proper to, and attainable only by the community, yet individually shared by its members" (Dupré 1993, p. 687). It was assumed that in normal situations the common good and the good of the individual would coincide. In case of conflict, the common good would be treated as the higher good.

Aristotle recognizes the conditions of the common good in both the virtuous character of the citizens and the institutional arrangements in the republic. Since the common good requires a shared life devoted to cooperative activities, it requires citizens to be just. Unfortunately most human beings are not just; they are perpetually engaged in competition for such external goods as honor, money or power. As there is always a scarcity of external goods, bad dispositions lead to intense competition and continuous conflicts. Consequently the main condition for arriving at the common good is the citizens' moral reorientation, which in turn implies education toward virtue.

Whereas in *Nicomachean Ethics* Aristotle teaches us that the common good cannot be achieved without the cultivation of the citizens' virtue of justice, in *Politics* he shows that it is essential to ensure constitutional arrangements which bind the good of the individual citizens to the good of the republic. Likewise it is imperative to involve all citizens in political deliberation concerning that good which is attainable by the community, yet individually shared by its members. Deliberation about our collective ends requires practical reason (*phronēsis*), which is developed through self-governance in one's household affairs and through participation in political deliberation (Aristotle 2002, pp. 1141b23–1142a30). Aristotle reminds us that at the heart of politics there lies "a quest to protect the integrity and political autonomy of each citizen in a political cosmos" (Terchek and Moore 2000, p. 911).

We think that Aristotle has convincingly shown that participatory self-rule is the highest political goal in itself; here the citizens' autonomy and the common good are bound together. If common good is not defined by means of reasoned deliberation about our collective ends, there is a great danger that under the guise of the common good, power groups such as businessmen, politicians or proponents of security may promote their private or sectarian interests. It is through participation in a reasoned, deliberative politics that we settle our differences and develop agreement about collective ends.

Let us now address the problems that occur when, without including people in the discussion, one value is chosen and set up to be higher than the others. For example, if without consulting the people it is decided that the primary value to be protected is security, in the name of which all other values are sacrificed, then the principle of moral autonomy is being violated. If violation of privacy is seen in the ordinary way as only the violation of individual rights, then what we see in reality is that limiting people's autonomy also results in damage to the functioning of democracy, which is required for the full self-realization of persons. Ruth Gavison has pointed out that "Privacy is also essential to democratic government because it also fosters and encourages the moral autonomy of the citizen, a central requirement of a democracy" (1980, p. 455).

## 7.3 Case Study of Second-Generation Biometrics Technology: Security *Versus* Privacy

To show that these theoretical ideas have practical relevance, we shall proceed to focus on the specific technology and illustrate the context of its implementations in two projects, INDECT and ADABTS, both funded by the Seventh Framework Programme of the European Union. In these projects, behavior monitoring and detection technologies are researched and developed in order to prevent terrorism and crimes in public places. One of the technologies used in these projects is second-generation biometrics.

In general, biometrics is a tool used to identify and reliably confirm an individual's identity on the basis of physiological or behavioral characteristics (or a combination of both), which are unique for a specific human being (FIDIS 2009a). Such characteristics include facial image, fingerprints, hand geometry, the structure of the retina or iris, DNA, gait, heart pulse, and voice. First-generation biometrical systems have been focused mainly on the question "who are you?", linking a person's different identities to his or her physical identity and thus serving the first aim of distinguishing one person from others.

New biometric technologies, called next-generation or second-generation biometric technologies, have a more ambitious aim: detecting "which person you are", based on an automatic interpretation or decision about the person, and resulting in a classification. The decision is made on the basis of some predetermined indicators of abnormal behavior that justify placing the person in a category of suspects who pose a potential risk or threat to the society (Sutrop and Laas-Mikko 2012, p. 22). Second-generation biometric systems are focused on intricate behavioral patterns, as indicated by gait or movement of the body, or by biological traits, states, and conditions of the body (e.g. heat, smell, ECG etc); using these patterns, the aim is to profile people on the basis of predictions of their actions and behaviors (McCarthy 2012).

Second-generation biometrics for security purposes are under development and testing; there are few available public materials about the results of such projects and their trial implementation, since these are kept confidential. EU has initiated some behavior detection research projects such as INDECT (Intelligent information system supporting observation, searching and detection for security of citizens in urban environments), ADABTS (Automatic Detection of Abnormal Behavior and Threats in Crowded Spaces), SAMURAI (Suspicious and abnormal behaviour monitoring using network cameras for situation awareness enhancement) etc. Let us proceed to a brief introduction to the INDECT and ADABTS projects. Our aim is to show, based on these examples, how second-generation biometrics technology implicitly relinquishes privacy and a host of other values in the name of security.

### *7.3.1 Intelligent Information System Supporting Observation, Searching and Detection for Security of Citizens in Urban Environment (INDECT)*

The main aim of the INDECT project[4] was to develop new, advanced and innovative algorithms and methods aiming at combating terrorism and other criminal activities, such as human trafficking and organized crime which affects citizens' safety (INDECT 2012a).

In order to achieve this aim, the project foresaw the development of the following items:

1. an intelligent information system for automatic detection of threats and recognition of criminal behavior or violence (with intelligent cameras);
2. tools for threat detection in the Internet (including the development of a new type of search engine which combines direct search of images and video based on watermarked contents etc);
3. techniques for data and privacy protection in storage and transmission of data.

The INDECT project encompassed threat detection in physical environments with intelligent cameras (streets, airports, football stadiums etc.) and in virtual

[4] The homepage address is the following: http://www.indect-project.eu/

environments (including computer networks, Internet, social networks such as Facebook, chatrooms); algorithms enable the automatic analysis and extraction of the 'abnormal behavior' that can advert to a possible crime or terrorist attack. According to the INDECT project homepage, the methodology aims, first, to detect specific crimes such as bomb attacks, robberies, Internet child pornography, and trafficking in human organs, and then to detect the source of the identified crimes (for example, specific criminals responsible for the crimes). Abnormal behavior definitions were not clearly articulated and proven in this project; some behavior patterns in physical environment such as forgetting luggage, rushing through the crowd, moving in the wrong direction, sitting too long in the airport, etc. were marked as suspicious (INDECT 2012b).

The prototyped information system had to integrate different algorithms and technologies for abnormal behavior and databases containing information for object recognition (car license plates) and the identification of persons. Automatic recognition of criminals and/or atypical situations and people, their parametric modeling and description, as well as systems for data processing and mining had to be developed and combined in order to build a large network-oriented security system that would assist security operators (INDECT 2012c).[5,6]

### *7.3.2 Automatic Detection of Abnormal Behavior and Threats in Crowded Spaces (ADABTS)*

The ADABTS[7] project aimed to facilitate the protection of European Union citizens, property, and infrastructure against threats of terrorism, crime, and riots by means of the automatic detection of abnormal human behavior (ADABTS 2012).

Similarly to INDECT, the ADABTS project was designed to enhance surveillance and security by exploring the possibilities for automated operator support. This was based on the imagery of abnormal behavior or unusual events, in which 'interesting' imagery can be distinguished. The extraction operations rely on signal-processing algorithms that detect predefined threat behaviors and deviations from 'normal behavior'. For real-time evaluation and detection of 'interesting' imagery,

---

[5] See also disputed video about the aims of the project, available at http://upload.wikimedia.org/wikinews/en/3/39/INDECT-400px.ogv

[6] The project was completed in 30.06.2014. It caused concern and attracted negative attention in 2010, when several European Parliament members were asking critical questions about personal data protection and transparency of the project results, for example E-1385/10, E-1332/10, E-2186/10, E-3190/10, E-1004/10 and E-6084/09. In 15.03.2011 the project was the subject of ethical review, which concluded that the research work is in full compliance with legal rules. After the debate the project homepage was improved and concentrated on answering questions about ethical issues. Participants confirm that no actual personal data are used for deployment of the system and several security controls (blurring photo images, watermarking the content, etc.) are used to protect video images.

[7] See homepage: https://www.informationsystems.foi.se/~adabts-fp7

data from audio and video sensors were to be combined with context information. Operators would be alerted only when suspicious cases are detected, and they would then consider further decisions on whether or not to take action. The abnormal behavior detection system would be used to monitor and provide additional guarantees for the security of certain events, or critical infrastructure such as international airports that are vulnerable to terrorist attacks and criminal activity (ADABTS 2012).

For our purposes it is important to understand what kind of imagery qualifies as 'interesting'. The aim of the ADABTS project was to arrive at such a definition; a list was created of indicators of abnormal behavior, as well as behavior models for scenarios in specific contexts: –large-scale events, crowded public spaces, and critical infrastructure. So far, distinct and visible behavior, such as whole-body behaviors (including movement within a space, excessive body gestures or gait), have been identified, as well as behaviors that are less obvious (such as signs of stress, eye movements, mumbling and sweating) (ADABTS 2011a, b). In keeping with the ideas of the ADABTS project, examples of abnormal behavior include rushing through the crowd, using an emergency exit, changes in heart rate, and recorded changes in body temperature. There are many more criteria and types of suspicious behavior and abnormal physiological indicators, which are listed in the section on project delivery (ADABTS 2011a, b). For security reasons, the participants in the first project did not make these public (Heck 2009); this material was published later. A final demonstration of the ADABTS system took place at ADO (Alles Door Oefening) Den Haag football club in June 2013.[8,9]

## 7.4 Ethical Values at Stake

Implementation of second-generation biometrics in security contexts is usually intended for massive surveillance (or dataveillance); this means not only the monitoring of specific suspects, but placing all people who happen to be in public places under surveillance and scrutiny. It can be argued, that "… activities performed in public are explicitly being made public by the individual performing them, because the person would have the choice of doing something different and knows that he or she can generally be observed by others in public places" (ADABTS 2010). On the one hand it seems that people are able to adapt their behavior when subjected to

[8] See press release from 05.07.2013 at homepage: https://www.informationsystems.foi.se/~adabts-fp7

[9] The project ended in July 2013. It did not receive a great deal of attention due to ethical issues. However, in the last demonstration Annemarie Sprokkereet from Ethical and Dual Use Advisory Board pointed out that deliverables of the project did not address the conceptual problems of abnormality and possible impacts in real life systems, nor the proportionality between the levels of perceived dangers and levels of profiling and surveillance, as well as many other ethical and legal issues (See slides of the last demonstration https://www.informationsystems.foi.se/main.php/ADABTS_Final_Demo.pdf?fileitem=7340169).

social control. However, we have to admit that these technologies are more powerful than old (man-powered) systems and that they are accompanied by many new risks. As shown by INDECT and ADABTS projects, second-generation biometrics is integrated into larger surveillance systems, making it easy to mine the data, profile or match it by combining different data sources, in this way obtaining additional information about a person. Biometrical data enables the creation of a profile for an identified person and the linking of other data to this profile. According to Helen Nissenbaum (1997; 2010), privacy in public places has to be protected, since in such cases of surveillance it is easy to transfer data from the context in which it was collected to another context, thus causing function creep.

The main ethical concerns about the application of second-generation biometrics concern issues of privacy, autonomy, and equal treatment. Since this technology is used to survey persons' behavior in secured areas and detect abnormal behavior and events, one result is the processing of huge amounts of personal data which are then collected into databases. Thus there are risks of data leakage or access by unauthorized persons, in both cases overriding the will of the data subject with respect to access and use of his or her data, consequently also violating his or her privacy. How can privacy be violated if data is collected anonymously? Although in most cases the data collected will indeed be anonymous (the focus is not on *Who you are* but on the question *What kind of person you are*), the problem is that it will still be possible to identify persons on the basis of comparing their video pictures with those in large databases, which already exist in several countries (e.g. in Estonia there are large databases of e-passport pictures).

Is this a reason for concern? On the one hand we might indeed feel more secure if new methods were available for detecting criminals and terrorists, thus proactively preventing attacks on our lives. On the other hand, there is an increased possibility of stigmatization and discrimination on the basis of false interpretation of biometric characteristics. Behavior prediction based on the collection of biometrics and identification may lead to the social classification and stigmatization of persons, automatically placing them in a category such as terrorist, criminal, unreliable or untrustworthy individual, etc.

In the case of second-generation biometrics, profiles are to be created about persons, and some people will be sorted out on the basis of different measurements of bodily behavior. Measurements form the set of the data that are 'mined' to detect the unique patterns for a particular person. Behavioral biometrics is the result of profiling, in which a certain kind of image is created and attached to the person, and then matched against data that can be used to provide more complete profiles (FIDIS 2009b). The main problem with profiling, besides data protection issues, is that it contains a stereotype of a possible offender, and this stereotype can inherit content from stereotypes of groups against which there is popular prejudice – and which is not evidence-based (Detecter 2008). According to David Lyon (2001), the surveillance involved in behavioral biometrics is a form of social sorting, of categorizing persons and groups, which accentuates differences and reinforces existing inequalities. We agree with Lyon that, unfortunately, these categories are seldom subjected to ethical inquiry or democratic scrutiny, despite their consequences for opportunities and choices in life. The reliability of these algorithms is under suspicion because of the high risk of a false error rate and a large number of fixed 'false images' of persons. "'Behavior' is a loose and socio-politically contingent concept," as Juliet Lodge (2010, p. 8) points out. She claims that "defining a certain type of behavior as deviant or indicative of 'risky intent' leaves behavior subject to the arbitrary interpretation, political vagaries, politico-ideological preferences and goals in power /.../." In this context, the following warning should be taken seriously: "Categories, descriptions and models are routinely imposed on individuals' identity information. We know what dramatic consequences the availability of labels like 'jew,' 'hutu,' 'tutsi,' and 'white,' 'black' and 'colored people' can have in administrative management systems for those concerned" (Manders-Huits and van der Hoven 2008, p. 2).

In addition to this problem of stereotyping through arbitrary interpretation of deviant or risky behavior, another essential feature of behavioral biometrics is that it allows on-the-move authentication or behavior identification. Traits such as the dynamics of facial expression or gait can be captured and analyzed covertly without any physical contact with the person and thus without his or her explicit knowledge and consent. Usually intentions are attributed to a person according to certain behavioral or physiological characteristics before this person has decided to do any harm. We agree with Manders-Huits and van der Hoven (2008, pp. 5–6) that this is problematic because it violates the person's right to be engaged in self-identification, the core of a person's moral autonomy. This argument follows from Bernard Williams's (1973) idea that respect for moral autonomy implies taking into account the other person's self-identification: we ought to understand the other person's aims, evaluations, attitudes, thoughts, and desires. In other words, if we assess the behavior of a person, we have to put ourselves in their shoes, taking into account his or her beliefs, motives and intentions, life projects, and related concerns. In the case of behavioral biometrics, identification of a person is performed from a third-person perspective without even attempting to interpret that person's motives. Thus, from the ethical point of view, the main problem with behavioral biometrics is that it does not make any attempt to take into account the person's self-identification; the person's behavior or physical characteristics are interpreted and viewed without any deeper knowledge of the person's own point of view. Thus, it may well be that the person's intentions or desires are misinterpreted.

It might be argued that today's world has become so dangerous that sorting people into categories is much less problematic than suffering from the threat of terrorism. Nevertheless, we should critically assess the proportionality of these measures. It has been pointed out that the risk of terrorism is overestimated, and that it is often manipulated in order to strengthen support for surveillance-based methods (Gray 2003).

Massive surveillance and dataveillance violate both privacy and autonomy; as we previously stated, these are not only individual values but also core values for a functioning democracy. Implementation of second-generation biometrics also harms other important pillars of democracy, such as the presumption of innocence and general social trust. Interestingly, experts in technical fields assert that behavioral biometrics is not invasive, since it does not require physical contact with persons. We find that capturing biometric characteristics from a distance is even more invasive. As one may not be aware of the fact that one's biometric features are being collected and analyzed, control over the processing of data becomes more and more difficult and everyday life begins to approximate a surveillance society – being watched by a Big Brother. The practice of large-scale surveillance causes a climate of general distrust in society and overall suspicion against everyone. These consequences are well known to those who have lived in totalitarian societies.

On the other hand, it is both understandable and natural that when risks to security are perceived to be increasing, people are ready to give up (some of) their privacy. We cannot overlook the fact that privacy is not an absolute value to be protected unconditionally. However, in each particular case, the restriction should be well founded and proportionate to the threat posed. As concerns biometrics, the conflict between individual and collective values seems genuine. As a common good, security is usually undisputable; otherwise the loyalty of the critic would be rendered questionable. The main problems that are overlooked are who makes the decision to collect biometrics in order to protect national security; how this decision is made; whether or not public discussion is enabled; whether or not all important stakeholders of society are to be included, and whether the implementation of biometrics as a security measure is proportional to the risks incurred.

What could we do in order to enhance security on the one hand and maintain privacy on the other hand? Usually it is suggested that privacy can be maintained by holding on to the requirement of informed consent and notification for the processing of the individual's data (Manders-Huits and van der Hoven 2008, p. 4). The failure to honor autonomy is expressed by the failure to obtain individual consent. In the case of behavior detection technology, it is not deemed possible to implement individual informed consent. As we have shown in our earlier article (Sutrop and Laas-Mikko 2012, p. 25), it is understandable that the procedures of informed consent are not implemented in contexts of national security, defense, and law enforcement. However, this does not mean that one should not respect people's autonomy. Granted, one really cannot make the implementation of safety measures voluntary, nor ask each individual for consent before collecting data. But it is still possible to inform people of the collection and processing of the data, as well as of the purpose of these activities; thus one offers an opportunity for control and a measure of standing up for one's rights. In addition, so-called public consent should be solicited in the use of technology. Even in the case of technologies where people are not aware of being under surveillance, their autonomy can be respected by allowing them, in other contexts, to participate in public discussion concerning the benefits and losses accompanying the implementation of such technologies, and in decision-making about whether or not such technology should be adopted.

In the projects discussed above, those implementing the projects have not analyzed the ethical and social consequences of the proposed technology. At least there are no materials available that would indicate such an analysis. There is an impending danger that once the technology reaches the implementation phase, there will be no more opportunities to analyze their possible influences nor to solicit public consent.

## 7.5 Conclusions

We have shown how important it is to understand the value of privacy. We began this article by showing that privacy is a mediating value, which protects other values. On this basis, we indicated that it is therefore dangerous to take the relinquishment of privacy lightly, even in the name of supporting security. We often do not notice that in this process we are relinquishing or failing to protect many other values, both individual and social, all of which are foundation stones of our society. Indeed, we desire to support security, but this cannot be done at the expense of destroying the basis of liberal democracy. We must at least be prepared to admit that security can be defended only in a completely different kind of society where there is no respect for human dignity, autonomy, and equal treatment, where trust is not honored, but replaced by total control and suspicion. Do we really want to live in such a society? Does this fit with our understandings of what is required for the flourishing of the human individual and living the good life? If not, we should avoid building a dichotomy between privacy and security and rather find ways how to strike a balance between these two important values, always taking specific socio-political contexts into account.

**Acknowledgements** We would like to thank Tiina Kirss for valuable suggestions and help with our English and Laura Lilles-Heinsar, Mari-Liis Tina and Tiina Randviir for careful assistance with our paper.

## References

Aristotle. 1996. *The Politics,* ed. Stephen Everson. Cambridge: Cambridge University Press.

Aristotle. 2002. *Nicomachean Ethics,* Trans. and ed. Roger Crisp. Cambridge: Cambridge University Press.

Automatic Detection of Abnormal Behavior and Threats in crowded Spaces (ADABTS). 2010. *ADABTS WP3 abnormal behaviour. D.3*. Available at https://www.informationsystems.foi.se/main.php/ADABTS_D3.1_Abnormal_Behaviour_Definition_Public_%28PU%29_final.pdf?fileitem=49119233. Accessed 17 Aug 2012.

Automatic Detection of Abnormal Behavior and Threats in Crowded Spaces (ADABTS). 2011a. *ADABTS WP3 abnormal behaviour. D.3*. https://www.informationsystems.foi.se/main.php/ADABTS_D3.1_Abnormal_Behaviour_Definition_Public_%28PU%29_final.pdf?fileitem=7340175. Accessed 7 June 2015.

Automatic Detection of Abnormal Behavior and Threats in Crowded Spaces (ADABTS). 2011b. *ADABTS WP2 user needs. D.2.1*. https://www.informationsystems.foi.se/main.php/ADABTS_D2%201_User_Needs_Final.pdf?fileitem=10010654. Accessed 17 Aug 2012.

Automatic Detection of Abnormal Behavior and Threats in Crowded Spaces (ADABTS). 2012. *Homepage*. https://www.informationsystems.foi.se/~adabts-fp7. Accessed 17 Aug 2012.

Detection Technologies, Terrorism, Ethics and Human Rights (DETECTER). 2008. *D25. Moral risks of preventive policing in counter-terrorism*. http://www.detecter.bham.ac.uk/pdfs/D05.1MoralRisksofPreventivePolicingv2.pdf. Accessed 9 Sept 2012.

Dupré, Louis. 1993. The common good and the open society. *Review of Politics* 55: 687–712.

Future of Identity in the Information Society (FIDIS). 2009a. *D3.10. Biometrics in identity management*. http://www.fidis.net/resources/deliverables/hightechid/#c2057. Accessed 17 Aug 2012.

Future of Identity in the Information Society (FIDIS). 2009b. *D7.12. Behavioural biometric profiling and transparency enhancing tools*. http://www.fidis.net/fileadmin/fidis/deliverables/fidis-wp7-del7.12_behavioural-biometric_profiling_and_transparency_enhancing_tools.pdf. Accessed 17 Aug 2012.

Gavison, Ruth. 1980. Privacy and the limits of law. *The Yale Law Journal* 89: 421–471.

Gray, Mitchell. 2003. Urban surveillance and panopticism: Will we recognize the facial recognition society? *Surveillance & Society* 1: 314–330.

Häyry, Matti, and Tuija Takala. 2001. Genetic information, rights, and autonomy. *Theoretical Medicine* 22: 403–414.

Heck, Wilmer. 2009. EU to monitor deviant behavior in fight against terrorism. *Spiegel Online*, 21 October. http://www.spiegel.de/international/europe/0,1518,656468,00.html. Accessed 2 Oct 2012.

Intelligent Information System Supporting Observation, Searching and Detection for Security of Citizens in Urban Environment (INDECT). 2012a. *Homepage*. http://www.indect-project.eu/. Accessed 17 Aug 2012.

Intelligent Information System Supporting Observation, Searching and Detection for Security of Citizens in Urban Environment (INDECT). 2012b. *D7.2 Creation of event model in order to detect dangerous events*. http://www.indect-project.eu/files/deliverables/public/INDECT_Deliverable_D7.2_v20100430_final.pdf/view. Accessed 17 Aug 2012.

Intelligent Information System Supporting Observation, Searching and Detection for Security of Citizens in Urban Environment (INDECT). 2012c. *D7.3 Biometric feature analysis component based on video and image information*. http://www.indect-project.eu/files/deliverables/public/deliverable-7.3/view. Accessed 17 Aug 2012.

Kupfer, Joseph. 1987. Privacy, autonomy, and self-concept. *American Philosophical Quarterly* 24: 81–89.

Lodge, Juliet. 2010. *Quantum surveillance and "shared secrets"*. A biometric step too far? Justice and home affairs. CEPS Liberty and Security in Europe. http://www.policypointers.org/Page/View/11489. Accessed 2 Oct 2012.

Lyon, David. 2001. Facing the future: Seeking ethics for everyday surveillance. *Ethics and Information Technology* 3: 171–181.

Manders-Huits, Noëmi, and Jeroen van der Hoven. 2008. Moral identification in identity management systems. In *The future of identity in the information society*, ed. Simone Fischer-Hübner, Penny Duquenoy, Albin Zuccato, and Leonardo Martucci, 77–91. New York: Springer.

McCarthy, Paul. 2012. Biometric technologies. In *Encyclopaedia of applied ethics*, 2nd ed, ed. Ruth Chadwick, 77–91. Amsterdam: Elsevier.

Moore, Adam. 2008. Defining privacy. *Journal of Social Philosophy* 39: 411–428.

Nissenbaum, Helen. 1997. Toward an approach to privacy in public: Challenges of information technology. *Ethics and Behavior* 7: 207–219.

Persson, Anders J., and Sven O. Hansson. 2003. Privacy at work – Ethical criteria. *Journal of Business Ethics* 42: 59–70.

Rössler, Beate. 2005. *The value of privacy*. Cambridge: Polity Press.

Solove, David J. 2007. "I've got nothing to hide" and other misunderstandings of privacy. *San Diego Law Review* 44: 745–772.

Sutrop, Margit, and Katrin Laas-Mikko. 2012. From identity verification to behavior prediction: Ethical implications of second generation biometrics. *Review of Policy Research* 29: 21–36.
Terchek, Ronald J., and David K. Moore. 2000. Recovering the political Aristotle: A critical response to Smith. *The American Political Science Review* 94: 905–911.
Travis, Alan. 2009. Fight against terror 'spells end of privacy'. *The Guardian*, 25 February. http://www.guardian.co.uk/uk/2009/feb/25/personal-data-terrorism-surveillance. Accessed 5 Oct 2012.
Williams, Bernard. 1973. *Problems of the self*. Cambridge: Cambridge University Press.

# Chapter 8
# Introducing Biometrics in the European Union: Practice and Imagination

**Kjetil Rommetveit**

**Abstract** This article gives a critical overview of the policy process introducing biometrics in the European Union. It starts out by describing how existing biometrics practices, such as fingerprinting and access control, are now being transformed and expanded in order to improve security and efficiency in the governance of global mobility. Following strong US initiatives in 2001 and onwards, the governance of mobility has been attempted transformed through what I call the biometrics vision. In brief, this vision states that biometric information can serve to increase control of mobility by enhancing the capacity of government agencies for monitoring and exchanging information about individuals. I follow this vision as it arose in the US, travelled through international organisations and into the European Union, where it has been promoted at high political levels. I finish off by a critical examination of the biometrics vision, referring to James Scott's concept of "seeing like a state", but using it in the context of the European Union. It is argued that the biometrics vision overlooks a number of technological and social issues. Social and technical complexities and problems may take on increased relevance as vision is turned into practice.

**Keywords** Biometrics • Imaginaries • Control • State vision • Regulation • European Union

## 8.1 Introduction

In its traditional form, biometrics means scientific measurements of the human body, and has been used in physiology for centuries. In this article, however, I will concentrate on its present-day, techno-scientific modality, which uses measurements for automated recognition of people. A typical biometric device increasingly encountered in everyday environments is the fingerprint reader. A number of other

K. Rommetveit (✉)
SVT, University of Bergen, Postbox 7805, Bergen N-5020, Norway
e-mail: kjetil.rommetveit@svt.uib.no



A. Delgado (ed.), *Technoscience and Citizenship: Ethics and Governance in the Digital Society*, The International Library of Ethics, Law and Technology 17,
DOI 10.1007/978-3-319-32414-2_8

devices and sensors are also becoming available, such as keyboard stroke, voice, iris scans and software-supported cameras, making possible recognition of faces, gait or behaviour from a distance. I shall focus on policy processes introducing large-scale biometric systems for controlling mobility in the European Union. As long since established for fingerprints, and more recently through DNA identification technologies, biometrics is inscribed with the notion that each individual embodies specific variations of universal bodily characteristics, and that these can serve to properly identify the individual. Furthermore, it is now presupposed that digital readers and information systems can use such unique characteristics to identify individuals and exchange their information for various types of social and political purposes.

I base my argument on the presupposition that present-day biometric applications emerge from well-known imaginaries about the socio-technical roles of information. *Information imaginaries* embody broad cultural representations about the character and roles of information, including how it can and ought to be used for social and technical purposes in the future. I shall focus on two different modes of imagining reality, these being simultaneously digital, communicative and physical:

*Digitalised Body Identities* It is a well-known fact of everyday life that people can be recognised according to their different bodily characteristics (with the face occupying a privileged position). This notion has long since been implemented into a number of socio-technical systems aimed at identifying and tracking persons in modern societies of strangers. Since its inception in the early twentieth century, fingerprinting has become widely known, first through criminal registries to keep track of criminals' stories, later through forensics as a way of linking perpetrators to crime scenes, but also for civil uses such as migration control (Cole 2001). Another well-known example is physical access control using hand geometry recognition, used in buildings and high-risk facilities since the 1970s. A US study has claimed the following as "the fundamental dogma of biometrics": "*An individual is more likely similar to him- or herself over time than to anyone else likely to be encountered*" (Pato and Millet 2010). However, how we interpret this statement must shift with the context. Systems of recognition have to work by singling out specific, and quite narrow, characteristics. Everyday recognition, on the other hand, performs an interpretative operation in which the person and her single characteristics, from visual appearance to character and intentions, emerge through constant oscillations of perception: from the whole to the part to the whole and back again. Increasingly, biometric information systems seek to mimic such judgements: single or multiple features (face, iris, fingerprints etc.) are registered and databases and registries are searched in order to identify the individual. When associated with traces of actions (such as travel and purchase) using software algorithms more complex images about individuals also emerge. In the so-called second generation of biometrics (see Sutrop and Laas-Mikko, Chap. 7, this edition) behaviour, is monitored and analysed in real-time, possibly creating a more complex image of character, plans and intentions. But it remains uncertain to what extent and with what consequences such "judgements" can be automated on a large scale.

*Interoperable Information Systems* The second mode of imagination on which I focus is of a more recent origin. It emerges out of a number of practices were increasing connectivity and sharing of information come to be seen as crucial to the fulfilment of important social and political purposes. Although a number of definitions are available, *interoperability* denotes the capability of different systems and operators to work together. In a number of applied engineering fields, from warfare to telecommunications to management, railroads and computer systems, the two terms "standardisation" and "interoperability" increasingly appear together: only where standards (such as digital codes, protocols, algorithms, and property rights) are shared across national, technical and regulatory boundaries, systems can be made to work together, in this case through exchange of information. Apart from positive connotations with increased networking in the age of the Internet, a main driving force for promoting interoperability may be seen as negative: it emerges from the realisation that most (information) systems do *not* work together and so enhanced connectivity and sharing of information becomes imperative.

I shall argue that information imaginaries (to be described throughout this text) are now being utilised and captured on by policy makers in efforts to re-imagine and re-create state sovereignty, especially as this relates to migration and border controls. Throughout the last 10 years, states and international organisations, including the European Union, have been working vigorously to implement globalised and globalising travel regimes (Salter 2006). In spite of the enormous political and technological resources that enter into such efforts, it remains to be seen to what extent existing practices and imaginaries can be re-cast as viable and functioning *practices* on a scale of much greater complexity, as required by the biometrics vision. In the following section I describe in some more detail important elements in and the genesis of this vision, before I turn to its reception in the EU.

## 8.2 A Transforming Vision

Shortly after 9/11 a great number of initiatives and new legislation were made by the US Congress to heighten security measures across a number of fields. The notion soon took hold that there is a strong connection between travel documents, the fight against terrorism and biometrics. The implications of the 9/11 Commission's statement that "for terrorists, travel documents are as important as weapons" (National Commission on Terrorist Attacks upon the United States 2004, 384), says something about the expanding scope of security in those days: extremely few people are terrorists, but almost everybody hold travel documents of one kind or other. The biometrics industry and lobby moved fast to ensure its place within this enlarged space of opportunity. The US Congress dealt with a great number of proposals for new legislation entailing biometrics and other security technologies (Zureik and Hindle 2004). One main outcome of the hectic legislative activity was the *US Patriot Act*. The *Department of Homeland Security* was set up to deal with terrorist threats and a *National Strategy for Homeland Security* was issued by the Bush administration. The ensuing *U.S. VISIT* programme require the taking of fingerprints and facial scans of all foreign nationals entering or exiting the United States, to be checked against databases and watch lists such as the *Terrorist Screening Database* and the AFIS system of the FBI (Aus 2006). By 2004 the Department of Homeland Security was demanding that all countries in the Visa Waiver[1] programme should implement biometrics in passports and travel documents by a set date (26 October 2004, but the deadline was later prolonged by 2 years). Countries that did not comply would be ousted from the programme.

Biometrics as imagined and enacted control underwent a radical up scaling and transformation within a very short timespan. What was retained, but also fortified and changed, was the basic reliance upon bodily information as unique markers of identity. In a testimony before the *US Senate Subcommittee on Immigration*, Executive Director Richard Norton of the International Biometrics Industry Association (IBIA) described how "biometric data is inherently secure and serves as a digital lock and key on personal information". The expression refers directly to a long-held scientific notion, stretching from early descriptions of the specificity of antigen to antibody, up until todays notion of genetic information and DNA fingerprints as specific to individuals. Just as the antibody refers to its antigen, so the digital image of body characteristics refers to the individual. The "lock and key" metaphor entails a notion of strong continuity between the body and the digital: Because bodies are made up of universal traits that appear in unique individual combinations, they become readable to biometric systems.

Biometric interoperability also received a powerful boost. The US National Security Strategy stated that "…government agencies storing terrorism information, such as terrorist "watch lists," have not been able to systematically share that information with other agencies. These differences can sometimes result in errors if, for example, visa applications and border controls are not checked against consistent "watch lists"." (US Office of Homeland Security 2002, 55). The document outlines a "system of systems" to deal with the integration of sources of information. As in many documents promoting interoperability, two elements are highlighted as crucial: first, information systems must be made to communicate. Second, legal and cultural barriers must be removed so that information can be exchanged freely. The implication is that biometric information may serve as a better medium for the connection of previously un-connected sources of information (mainly databases), and that this will turn out beneficial in operational and organisational terms. In the case of passports, these already share a machine-readable zone containing basic information about the holder (name, passport number, nationality, date of birth, sex, expiration date and personal identity number). However, this information is based on alphanumeric data, not biometrics, and increasingly considered a security liability.

What was radically upgraded was the scale and complexity of operations: According to Wikipedia, in 2007 the US VISIT programme recorded 46,298,869

[1] Citizens of countries included in the US Visa Waiver program can travel to the US for up to 90 days without a visa.

entries at air and seaports. In the case of existing practices such as automated fingerprint systems (AFIS) and Physical Access Control, operators can rely upon highly controlled circumstances for the taking of fingerprints. Employees or others to be granted access to a building typically comply with the process of enrolling in the biometric database; criminals are made to comply during detainment and under threat of force. But how do you apply biometrics to the setting of a highly busy airport?[2] How do you get millions of people to collaborate in the enrollment of biometric data? And how do you identify and detain a terrorist based on fingerprints, if those fingerprints do not already exist in a database? Analysts and industry alike gave the answer, as in John Woodward Jr.'s highly suggestive article title *Biometrics. Facing up to Terrorism*, and the company Visionics white paper *Protecting Civilization from the Faces of Terror*. As argued by Woodward, *"While fingerprint and other biometric systems could be used to identify these individuals, government authorities might find it difficult to collect the fingerprints or iris scans of suspected terrorists in order to build the database against which to compare an unknown individual. Facial recognition biometric systems, however, offer a way around this problem. Specifically, facial recognition systems will allow the identification of a suspected or known terrorist even if the only identifying information we have is a photograph"* (Woodward 2001, 10).

Whereas it was noted that the technology was still far from perfect, the author nevertheless claimed that research and industry would soon pick up the challenge and improve systems functionality. This illustrates how, in the field of large-scale biometrics there are good reasons to claim that imaginations of *future* performance, and not present results, are constitutive of large socio-technical experiments. Woodward also claimed privacy reasons in favour of facial recognition over fingerprints, as privacy regulations or the US Constitution did not cover the taking of automated facial images in a crowd. Hence, the loophole in privacy regulations opened up by increased technological capacity was seen more as an option, than as a problem.

The expanded biometrics vision was exported mainly through two international organisations: the G8 and the International Civil Aviation Organisation (ICAO). The G8 provided a forum in which the leaders of the industrialised world agreed to introduce the technology; the ICAO worked out technical specifications in order to promote standardisation and interoperability on a global scale. The 2002 Berlin Resolution of the ICAO decided on the *face* as the globally interoperable standard for passports and travel documents. Following this, the main document has become the ICAO *Doc. No 9303*, making the facial image the primary and mandatory biometric, with fingerprints and iris scans as optional alternatives.

[2] For the European setting, the problem is even bigger: estimates are that a total number of 300 million people travel in and out of the region annually (EC 2008).

## 8.3 Introducing Biometrics in the EU

The biometric transforming vision easily lent itself to political visions of enhanced border control as a way of promoting European integration. The relevance of 9/11 for increased use of biometrics was openly recognised by EU policy makers: *"In the aftermath of the tragic events of September 11, 2001 the Commission was asked by Member States to take immediate action in order to improve document safety"* (European Commission 2003a). Notably, the "Member States" in question were mainly those taking part in the G8: the UK, Germany, France and Italy (also joined by Spain). From the outset, a unified and overarching approach was pursued. The 2003 Council of Thessaloniki stated that: *"...a coherent approach is needed in the EU on biometric identifiers or biometric data, which would result in harmonised solutions for documents for third country nationals, EU citizens' passports and information systems (VIS and SIS II). The European Council invites the Commission to prepare the appropriate proposals, starting with visas, while fully respecting the envisaged timetable for the introduction of the Schengen Information System II"* (European Council 2003). This constitutes the EU parallel to (and continuation of) the US biometrics vision where, again, the two initial information imaginaries are conjoined in a (seemingly) unitary vision.

But it also marks a new stage in the project of European integration and the wider contexts of that process. This was not only so because of strong calls for tighter security measures in controlling the common external border; it also came along with the eastward expansion of the EU towards ten new member states in 2004. Although originating from the same holistic vision, ensuing legislative measures broadly followed the two "biometric imaginaries" sketched above: on the one hand promoting the tightened relation between the body and travel documents, on the other, the tightened integration of systems and government agencies as given through the concept of interoperability. These would issue in two coordinated, still separate, batches of legislative initiatives, to which I now turn.

*Biometrics in Travel Documents* In September 2001 the European Commission submitted proposals to the Parliament and Council for enhancements of security standards in visas and residence permits for third country nationals, both of which were adopted in February the following year. Neither proposal included biometrics. Following the Councils of Laeken, Sevilla and Thessaloniki, however, the proposals for visas and residence permits were again amended, this time with the twin aims of bringing forward the implementation of security standards (from 2007 to 2005), and to introduce biometric identifiers. In September 2003 the Commission issued another proposal (European Commission 2003a) for amending the regulations of visas and residence permits. This time facial photography was included as the primary biometric identifier and fingerprints as secondary. Both were obligatory and both were to be implemented on the medium of a contactless chip.[3]

[3] Due to technical problems with the visa stickers, the proposal was not implemented into law until 2006. The choice of contactless chips entailed the use of RFIDs, thus settling for a more compli-

Following this, and using the same technical committee as for visas and residence permits, parallel proposals were developed for biometrics in European citizens' passports. In February 2004 the Commission presented a proposal, the main rationale of which was to "establish a reliable link between the genuine holder and the document" and so to "fight the use of false documents" (European Commission 2004). More reliable documents, it was surmised, would also have important operational advantages, since "bona fide citizens will pass more smoothly through border controls" and "those who use forged or fraudulent passports will have less chance to enter the territory of Member States" (*ibid.*). In contradistinction to visas and residence permits, it was proposed that only the facial image be made obligatory. More controversial was the promise of a centralised, biometrics-based EU passport register, which prompted a number of sharp reactions, from both data protection officials and activists. However, the final regulation went further than the initial Commission proposal: whereas the prospect of a centralised biometric database was maintained, the Council also included the second biometric, the fingerprint, as obligatory (Council Regulation 2004). The reason for including the second identifier (not obligatory according to ICAO Doc 9303) was the enabling of one-to-many searches in existing fingerprint databases (such searches would not be possible on the basis of the state of the art of face recognition technologies). The Council decision, however, was far from unanimous and came out of a private meeting of the "G5"[4] in Florence, Italy (La Repubblica 2004), and so by-passing ordinary decision making procedures in the Council (Aus 2006).

The process of introducing biometrics into travellers' documents in the European Union came as the result of concerted processes in high political circles. Decisions were not made without opposition, but criticism never ventured far outside elite levels, such as ministers from smaller countries in the Council, privacy commissioners and privacy advocates. The European Parliament arranged for a hearing, including technical and legal expertise. The ensuing report criticised the lack of democratic control and the mis-match between political and technical decision making: *"It should be emphasised that the European Council made a political decision to introduce biometric identifiers in EU passports without any input from practitioners and without knowing the magnitude of the problem"* (LIBE 2004).

The main reason given for biometrics in travel documents remained that of establishing "a more reliable link between the passport holder and the travel document". In terms of control, this implies a movement away from face-to-face checks (by border guards, visa and immigration authorities), towards an increased emphasis on two mutually constituting poles: the digital and the body. But this does not fully account for how biometrics was supposed to contribute to "the fight against terrorism". Almost as if by necessity, then, the anti-terrorism agenda and the biometrics vision push for further technological measures to be implemented, measures that go

cated option than that chosen in the US. The inclusion of RFIDs in travel documents and passports has been a source of much criticism, not the least due to increased risks of spoofing (i.e. hacking into documents from a distance).

[4] The 4 European G8 member states + Spain.

beyond the program of "improving document safety". These measures intensify digitalisation, automation and interconnection and introduce further elements into operations of checks and controls.[5] This directly leads onto the interoperability imaginary and its related legislative initiatives.

*Interoperable Information Systems* As required by the Council at Thessaloniki, the biometrics strategy was seen in conjunction with the establishment of a number of large-scale biometric information systems.[6] The example *par excellence* of such a system would be the SIS and its transformation into SIS II.[7] From the outset, the purpose of SIS ("the backbone of Schengen") was to "maintain public order and safety" (CIS Art 93) by fortifying controls and security at the external Schengen border as the internal borders were abolished.[8] The system contains alerts on people (and property) to be extradited, denied entry, placed under surveillance or interrogated. The impetus for expanding the original system, operative since 1995, was not to fight terrorism, but came as a natural result of the expansion of the EU. The need for technological upgrades was also important, but biometrics was not mentioned. Two years later, this had changed: *"the system should provide the flexibility to incorporate new functionalities, as well as new information and rules without major technical changes. This would include the inter-linking of alerts and the use of biometric information"* (European Commission 2003b). Another important concept had also found its way into the document, that of *interoperability*. On a conspicuous level, this entailed the requirement that the member states implemented compatible (or "harmonised") standards for the exchange of biometric data within the SIS II itself. But it also came along with the further requirement that *"The compatibility with other relevant – existing or future – databases in these fields is of the utmost importance"* (*ibid.*). This entails that the central system, to the greatest possible extent, be made interoperable with existing national systems (especially automated fingerprint identification systems, AFIS). But it also meant that the SIS II be made to exploit potential synergies with other European systems, the VIS, but also with the EURODAC system (European Commission 2005).

Whereas the SIS II had been planned since the 1990s, the Visa Information System was a genuine child of the situation in the early 2000s. The idea of a centralised system for the collection and exchange of visa data was put forth by the German government in the direct aftermath of 9/11 (Aus 2006). A Commission feasibility study estimated that the system would connect the visa authorities of 27 countries, 12,000 operators and 3500 consular posts worldwide (European Commission

---

[5] Even more radical measures include efforts to develop systems that may detect "malignant intentions" and suspicious behaviour, see Sutrop and Laas-Mikko in this edition.

[6] A number of other databases, not dealt with in this article, are also being developed. VIS SIS II and Eurodac are among the biggest and most important.

[7] Denoting, respectively, the first and second generation of the Schengen Information System. The plan was for transition from SIS to SIS II in 2007, but rollout has been delayed due to a number of problems.

[8] Some of the rationale of this is further described by Gunnarsdottir, this edition.

2003b). The system was anticipated to process approximately 20 million applications each year. The data would be stored for 5 years, resulting in a number of approximately 70 million datasets at any given time. Awaiting the possible implementation of the European Passport Registry and the SIS II, this would make the VIS the largest biometric database in the world. The system will share a "common technical platform" with SIS II. It consists of a centralised base, C-VIS, operating and coordinating the system of national contact points, called N-VIS, and these will be connected to local contact points, such as consulates or immigration offices. The systems would also share the Biometric Matching System (BMS), required for reading and comparing biometric data.

The two systems should have separate legal bases, and their uses kept separate, in the sense that data stored on SIS II should not be matched with data stored on VIS (or EURODAC, or any other system). The main users of SIS II would be police, border guards, internal security and immigration authorities; for VIS it would be consular posts, border guards and immigration authorities. Cross-uses were imagined, for instance by giving consular posts access to SIS II data as part of visa procedures or internal security access to visa or immigration data. Consular and immigration authorities already have access to the SIS, and they will be given continued access to the SIS II. What was new was access for police, security and judicial authorities to the VIS. For such access to be granted, extraordinary circumstances would have to apply, i.e. "overriding public security concern" (European Commission 2005). But in spite of such precautions the purposes of the systems are expanded under this vision. The SIS II is no longer restricted to "security checks" (as with the SIS) but has become a system for "prevention, detection or investigation", potentially including alerts on "persons who are likely to commit serious offences" (*ibid.*). The interlinking of alerts has the potential of creating new kinds of information on individuals (such as profiles). Similar things go for the promise to use biometrics and not alphanumerics for database searches (a facial image or fingerprint may generate much larger amounts of matches from a radically expanded set of sources). Finally, the number of agencies to be granted access has expanded, and the purposes of the systems are defined in looser terms in order to remain "flexible".

In the 2005 Hague program the described developments were included within an "integrated management" of the external borders (initiated at the Laeken Council in 2002 and pursued in the Stockholm programme and Lisbon Treaty). Following the Hague program the Integrated Border Management Strategy has been fortified and now makes up the most comprehensive framework for understanding and implementing biometrics for migration control in the Schengen area (European Commission 2008). Through interoperable biometric systems the border is imagined as fortified and enhanced in at least two respects. First, the border itself is enhanced through increased capacities of border guards for "seeing" and controlling individuals at the same time as throughput is enhanced through automation: *"One border guard should be able to oversee up to ten automated border gates in operation. Automated border controls for bona fide travellers would provide major benefits in time savings on crossing the external border and allow border authorities to focus their resources on those groups of third country nationals that require more attention, thus improving overall security at borders"* (*ibid.*). Second, the border itself is expanded in time and space, even to the extent that "the border is everywhere" (Lyon 2005). The new management strategy entails "measures taken at the consulates of third countries, measures at the border itself and measures inside the Schengen area" (European Commission 2008). At consulates, visa applicants will be subject to a more thorough pre-screening process, by being checked against VIS and SIS II. Awaiting the rollout of VIS (from 2009 onwards) an entry-exit database was also envisioned that could keep track of visa over-stayers (by far accounting for the greatest amount of "illegal immigrants" in the EU).

## 8.4 Seeing Like a Union?

At this point I want to add James Scott's concept of "seeing like a state" to the analysis, and so begin to look for some ways to assess the wider implications of the biometrics vision. Scott describes how a certain mode of state vision has been endemic to the increasingly specialised and expert-dominated planning and organisation of modern societies. It is targeted and purpose-directed, constructing and projecting a kind of map "designed to summarize precisely those aspects of a complex world that are of immediate interest to the map-maker and to ignore the rest" (Scott 1998, 87). Biometrics is literally a technology that provides what Scott terms "eligibility": it renders subjects and citizens visible to the state, hence also (potentially) controllable (Lyon 2009). In the case of biometrics, security technologies and mobility control, previous US Secretary of Homeland Security, Michael Chertoff, has succinctly described the biometrics vision and its related interest thus:

> Allow me to share with you where I would like to see us move – toward a world that is banded together by security envelopes, meaning secure environments through which people and cargo can move rapidly, efficiently, and safely without sacrificing security or privacy... For those within the security envelope, we will have a high degree of confidence and trust, so that trusted travellers and shippers don't have to be stopped at every point along the way to be re-vetted and rechecked. And that would enable us to focus more of our resources for those outside the security envelope – for the kind of in-depth analysis and the kind of in-depth vetting that is necessary to make sure those who seek to harm us do not slip through the cracks (US secretary of Homeland Security Michael Chertoff speaking on US – EU relations, Berlin 2005).

The concerted pushing of Chertoff's "security envelope" on high political levels structures the kinds of policy responses to be considered as the technology becomes implemented. Although still early days for assessing the wider social implications of biometric information systems, certain lessons may be drawn from looking at the kinds of policies that are enabled by this vision. Although focused on the creation of a more secure global mobility system, the overall vision constructs and imagines it's own implementation in specific ways. It is inscribed with implicit presuppositions about human and technological agency necessary for its own realization. It is not sure, however, that these presuppositions are representative of how things and people behave in practice.

First of all, this pertains to how users (citizens and officials meant to operate the systems), are imagined. Clearly, for large-scale biometric systems issues of public perception and acceptability cannot be ignored: without subjects' cooperation they will not work. However, acceptability is predicated on the prior acceptance of the over-all vision: In order for acceptance and enlightened debate to take place, society must be educated. This was clearly articulated in a joint EC/industry report that tried to assess the experiences with the implementation of large-scale biometric systems in Europe to-date (i.e. 2008):

> There is a need for initiatives leading to widespread public awareness amongst EU citizens as to the purpose and use of biometric technologies in large schemes such as e-passport and public administration applications. If the purpose of the system is clearly explained to the citizen, and also the way the citizen is expected to interact with the system, and if the safeguards are in place with their resulting benefits, all stakeholders involved would be in a better position to understand their role in biometrics deployment. A fair and open debate could then commence with discussions on costs/benefits, purpose of systems, potential impacts, in the long run ensuring system take-up and use, (Goldstein et al. 2008).

However, between the level of public acceptance and high-level policy making there is another (essential) level of actors that needs to be taken into account, namely those intended to implement and operate the systems: engineers and software developers, but also officials such as immigration officers, border guards and law enforcement agencies. A main issue with interoperable systems is that these groups of operators and their related institutions should be made to communicate, exchange information and collaborate in ways that may move societies closer to Chertoff's "world banded together by security envelopes". However, in practice, and especially in the EU, serious problems emerge as operators are expected to collaborate across national legislatures, operational cultures and borders. In 2006 the Commission diagnosed the following problem relating to the implementation, testing and operation of systems: *"...wider and more direct consultation with Member States and exchange of best practices would be useful...more consistent introduction and use of certain data...should be made by Member States"* (European Commission 2005). Problems relating to such *lacking exchange of information* were repeated in a 2009 communication from the Council. At this point in time, the SIS II implementation was in a state of deep crisis, to the extent that an alternative plan was developed for the eventual case that the initial system had to be abandoned. Although reasons for the troubles suffered by the system were complex, a key issue identified in the *SIS II analysis and repair plan* was the low level of participation among member states in testing the system (European Council 2009). Similar issues were emphasised in the already mentioned 2008 EC/industry report. Crucial problems were encountered during attempts to map public authorities experiences with biometric systems, most of which were "citing confidentiality reasons" for their lack of collaboration (Goldstein et al. 2008). Serious problems arise as industry and the Commission try to introduce interoperable systems in an area highly sensitive to national states:

> One overall recommendation for EU policy-makers is to create consensus among the Member States and implement a procedure that would facilitate the open dissemination of all information on biometrics systems in the implementation phase. This lack of information concerning EU Member States large-scale projects does not bode well for biometrics deployment in the future as keeping this data secret could suggest that the systems are not secure, may hide poor error rates, be behind schedule or conceal unsatisfactory roll-out results (ibid.).

Consensus as to the use and purpose of systems is a matter to be achieved. As seen, however, the consensus in question was created by a small number of policy makers on high levels. How, then, can one expect consensus prior to systems implementation? The very point of introducing biometric systems is to improve exchange of information and collaboration between government agencies and officials. Trust and collaboration were imagined according to a simplified vision of communication. Also downplayed were the technical complexities of biometric systems. In the words of researchers who have been closely involved with the process of introducing e-passports in Holland and in the EU:

> The effectiveness of biometry is highly overrated, especially by politicians and policy makers. Despite rapid growth in applications, the large-scale use of biometry is untested. The difficulty is that it is not only unproven in a huge single application (such as e-passports), but also not with many different applications in parallel (including "biometry for fun"). The interference caused by the diversity of applications—each with its own security policy, if any—may lead to unforeseen forms of fraud (Hoepman et al. 2006, 154).

The biometrics vision, then, presupposes and projects technological models of human and social agency and rationality onto its users (according to Brian Wynne a prominent tendency of many science and technology visions). It cannot be known in advance that social responses will be "rational" in the sense here presupposed. This is especially so for dual use technologies used for multiple purposes and by multiple agencies, and especially those that contain clear surveillance characteristics.

Similar issues pertain to the conditions for ethical and legal debate: also here the biometrics vision prescribes certain solutions whereas leaving others out. The predominant metaphor in regulatory debates has become that of a trade-off model, in which the governing metaphor is that of "balancing" security with freedom and privacy (Balzacq and Carrera 2006). In the 2008 policy report a pertinent question is asked: *In what way, and based on which options, are decisions made when identifying an appropriate level of trade-off that involved citizens would have to make in relation to values that may need to be sacrificed?* (Goldstein et al. 2008, 11). Because the over-arching vision was simply taken for granted, the authors also took for granted that certain values, i.e. privacy and freedom, will have to be sacrificed in order for biometrics systems to have the desired effect, i.e. to improve security and efficiency in the control with mobility. The required acts of balancing presuppose that biometric systems work as projected, and that they are received and perceived accordingly by their users and the wider public. This, however, should not be taken for granted if "a fair and open debate" is to be had about biometric information systems.

To return to Scott's concept of "seeing like a state": his was an analysis of how "certain schemes to improve the human condition have failed", especially those instigated to promote centralised planning and control. What is also intrinsic to the projects described by Scott is that they fail because they ignore their own environmental and cultural contexts and preconditions. It cannot be known that this will be the destiny of biometric systems in the European Union. But they do seem to share important characteristics with the projects described by Scott.

## 8.5 Conclusion

It is possible that biometrics may actually improve the security of travel documents, passports or other processes requiring verification of identity. Still, in this article I have described and criticized the emergence of a certain biometric vision, used to implement and direct large-scale biometric systems in the EU and beyond. I have argued that this vision is overstretching in terms of technological promise whereas ignoring social implications. For such reasons it is likely to lead to unforeseen difficulties and complexities in practice. It is difficult to predict how biometric systems will work to influence our societies in the years to come. Dominant visions and the policies informed by them, as here described, should be placed under greater public scrutiny. Greater attention to the concrete workings of biometric systems and their interactions with groups, individuals and societies will be needed for more democratic debates, policies and technology applications.

**Acknowledgements** This paper was made possible by the research opportunities emerging from the FP7-funded project TECHNOLIFE (project number 230381). The author also thanks Statewatch, who made available many of the documents cited in this article.

## References

Aus, Jonathan. 2006. *Decision making under pressure: The negotiation of the biometric*. Oslo: ARENA.

Balzacq, Thiery, and Sergio Carrera. 2006. *Security versus freedom? A challenge for Europe's future*. Aldershot: Ashgate.

Cole, Simon A. 2001. *Suspect identities: A history of criminal identification and fingerprinting*. Cambridge, MA: Harvard University Press.

Committee on Civil Liberties, Justice and Home Affairs (LIBE). 2004. Report on the Commission proposal for a Council regulation amending Regulation (EC) No 1683/95 laying down a uniform format for visas and the Commission Proposal for a Council regulation amending Regulation (EC) No 1030/2002 laying down a uniform format for residence permits for third-country nationals. European Parliament, Brussels.

Communication from the Commission to the European Parliament, the Council, the European Economic and Social Committee and the Committee of the Regions. 2008. Preparing the next steps in border management in the European Union. European Commission, Brussels.

Council Regulation (EC) No 2252/2004 of 13 December 2004 on standards for security features and biometrics in passports and travel documents issued by Member States. European Council, Brussels.

European Commission. 2003a. Proposal for a Council Regulation amending Regulation (EC) 1683/95 laying down a uniform format for visas. Proposal for a Council Regulation amending Regulation (EC) 1030/2002 laying down a uniform format for residence permits for third-country nationals. European Commission, Brussels.

European Commission. 2003b. Communication from the Commission to the Council and the European Parliament. Development of the Schengen Information System II and possible synergies with a future Visa Information System (VIS). European Commission, Brussels.

European Commission. 2004. Proposal for a Council Regulation on standards for security features and biometrics in EU citizens' passports. European Commission, Brussels.

European Commission. 2005. Communication from the Commission to the Council and the European Parliament on improved effectiveness, enhanced interoperability and synergies among European databases in the area of Justice and Home Affairs. European Commission, Brussels.

European Commission. 2008. *New tools for an integrated European Border Management Strategy*. MEMO/08/85. Brussels.

European Council. 2003. Presidency Conclusions – Thessaloniki, 19 and 20 June 2003. Retrieved from: http://www.consilium.europa.eu/uedocs/cms_data/docs/pressdata/en/ec/76279.pdf

European Council. 2009. Implementation of the February JHA Council conclusions on SIS II. Note from the Presidency to the Article 36 Committee. European Council, Brussels.

Goldstein, James et al. 2008. "Large-scale biometrics deployment in Europe: Identifying challenges and threats". European Commission, Joint Research Centre Institute for Prospective Technological Studies, Seville.

Hoepman Jaap-Henk et al. 2006. Crossing borders: Security and privacy issues of the European e-Passport. In *Advances in information and computer security*, LNCS, ed. H. Yoshiura et al., 152–167. New York: Springer.

International Civic Aviation Association. 2006. Machine readable travel documents. http://www2.icao.int/en/MRTD/Pages/Doc9393.aspx. Accessed 27 July 2011.

La Republica 18 October 2004. Impronte sui passaporti nella UE.

Lyon, D. 2005. The border is everywhere: ID cards, surveillance, and the other. In *Global surveillance and policing: Borders, security, identity*, ed. E. Zureik and M. Salter, 66–82. Cullompton: Willan.

Lyon, David. 2009. *Identifying citizens: ID cards as surveillance*. Cambridge: Polity Press.

National Commission on Terrorist Attacks upon the United States. 2004. The 9/11 Commission report: Final report of the national commission on terrorist attacks upon the United States. http://www.9-11commission.gov/report/index.htm. Accessed 27 July 2011.

Passport in the Council. Arena working papers series No. 11. University of Oslo.

Pato, Joseph N., and Lynette I. Millet. 2010. *Biometric recognition: Challenges and opportunities*. Report for the National Research Council. Washington, DC: The National Academies Press.

Salter, Mark B. 2006. The global visa regime and the political technologies of the International self. *Alternatives: Global, Local, Political* 31(2): 167–189.

Scott, James C. 1998. *Seeing like a state: How certain schemes to improve the human condition have failed*. New Haven/London: Yale University Press.

US Office of Homeland Security. 2002. US National Security Strategy. http://www.au.af.mil/au/awc/awcgate/nss/nss_sep2002.pdf. Accessed 27 July 2011.

Woodward, John D. 2001. *Biometrics: Facing up to terrorism*. Santa Monica: RAND, Arroyo Center.

Zureik, Elia, and Karen Hindle. 2004. Governance, security and technology: The case of biometrics. *Studies in Political Economy* 73: 113–137.

# Part IV
# Governing Spaces

## On Space Technologies and New Forms of Visibility

# Chapter 9
# Digital Globes: Layers of Meaning and Technology, Redefining Geographies and Communities

**Kjetil Rommetveit, Ângela Guimarães Pereira, and Tiago Pedrosa**

**Abstract** This chapter aims to give an overview of some developments in digital maps and globes in the last decade, and some ways in which we, individually and collectively, experience and imagine ourselves, others and space through the use of digital maps and globes. We will focus on modes of imagination as ways of understanding collective experience, and overlaying as the technological and digital counterpart needed for constructing such experience. We also offer some examples of ways in which collective imagination is self-replicating and co-produced along with broad technological and social platforms, possibly also changing communities or generating new forms of community. Anything like a total overview is impossible, and so our selection is an eclectic one, in both diachronic and synchronic terms. The exposition is based in the work with one European research project, Technolife. During that project we have been reviewing academic as well as policy literature on GIS, and we have been carrying out a debate about the social and ethical aspects of GIS with a number of online participants (www.technolife.no). In an essayistic style we draw upon all these sources, including media and web content as well as arguments made by participants during the online debate. In order to introduce our argument, the first section will give a brief account of how, in time, geographical space has been codified into disciplines and political enterprises, thereby also feeding into different communities and world-views.

**Keywords** GIS • Ethics • Dialogue

K. Rommetveit
SVT, University of Bergen, Postbox 7805, Bergen N-5020, Norway

Â. Guimarães Pereira (✉) • T. Pedrosa
European Commission – Joint Research Centre, Ispra, Italy
e-mail: angela.pereira@jrc.ec.europa.eu



A. Delgado (ed.), *Technoscience and Citizenship: Ethics and Governance in the Digital Society*, The International Library of Ethics, Law and Technology 17,
DOI 10.1007/978-3-319-32414-2_9

## 9.1 Imagining and Constructing Space

Since the beginning of modern science, space has made up a distinct dimension of human fascination, imagination and even obsessions; the near and the remote, the physical and the virtual, the visible and invisible, utopian and dystopian. These concepts, and the phenomena constructed by them have been continuously interrogated, disciplined and re-invented. Space has been codified in many disciplines, the most obvious being geography, astronomy, philosophy, mathematics and physics amongst others. According to the online Britannica Encyclopedia (2011a), space is a "boundless, three-dimensional extent in which objects and events occur and have relative position and direction". The concept of space as a neutral and objective entity was central to the emergence of the modern age. The idea was first made prominent by Galileo's natural philosophy, where physical space was depicted in its "primary qualities" such as extension, length and depth; these were seen as totally different from, and altogether more reliable than human experience and sensibility. Perhaps paradoxically, motivations to construct space as objective and neutral have lead to the development of enterprises of a highly political nature, such as mapping and cartography.

Cartography is an ancient discipline that dates from the prehistoric depiction of hunting and fishing territories (Britannica Encyclopedia 2011b). Early in the last millennium it was an art of representing knowledge about places in graphical form. But the Earth was not always depicted as a sphere; the coordinates were not always represented the way they are now; scales were often not related to the size of the territory but to its power, etc. It is only with navigation in the fourteenth century that more accurate geographical representation started. Maps were then seen as political instruments, even secrets of state, powerful tools to explore the world. They represented a competitive advantage for exercising power and control over others.

As maps became accessible to and appropriated by ordinary citizens they became intrinsic to the ways in which we imagine our place in the world: how we see ourselves in the larger geopolitical and cultural scheme of things, but also how we organise private life. Home atlases, geography teaching in schools, sea navigation or hiking maps and weather broadcasts appear to be disinterested forms of representing spatial features. At the same time we also connect with them on intimate levels of identity, culture and belonging: the shape of our countries merge with the colours of national symbols like flags, as well as with other important cultural and natural icons.

The last century saw tremendous evolution in the ways maps were produced. For example, Mercator projection maps (sixteenth century), which increase disproportionally areas close to the poles, were abandoned. Still in the late 1980s and 1990s, the American Cartographic Association was discussing which were the best and most suitable projections systems, and urged book and map publishers, the media and government agencies to cease using rectangular world maps, as they promoted erroneous conceptions of the world (Robinson 1986). Aerial photography increased the accuracy of topography and morphology of the terrains, digital technologies provided more precision in geo-referencing and overlaying – a development connected to Computer-Aided Design (CAD) and Geographic Information Systems (GIS). In the next section we describe the rise of the digital in mapping and how the concept of *overlaying* became the predominant tool (and skill) for depicting and imagining space in the digital age.

### *9.1.1 The Rise of the Digital*

In the late 1960s Waldo Tobler, an American geographer, developed a simple model that tied together computers and geography. A "map in-map out", the so called MIMO structure allowed maps to be available in a digital format, which then could be manipulated in order to produce new maps. This system is generally classified as "computerized cartography", and set the stage for the development of GIS (Geographical Imaging Systems) by Roger Tomlison, generally considered to be the father of these types of systems (in Encyclopaedia Britannica[1]). GIS can be described as a computer aided system which provides components for acquisition and conversion of spatial data into a digital format; a storage component of the digitised spatial data; an analysis module, containing algorithms that perform different types of operations with the geo-referenced data; and a system to display and produce diverse outputs.

In GIS we can distinguish two main types of maps: general-purpose maps, which contain different spatial features and thematic maps which focus on a specific topic or spatial feature such as population, soil type, housing, etc. The latter are more commonly used in GIS.

The system of overlays (or more commonly designated as *overlaying*) is one of the keywords in GIS. Overlaying is about combining different types (layers) of information into a single output. Although for a long time part of mapping techniques, GIS made it explicit, easy to use and increasingly automatic. This relatively simple yet versatile technique allowed the simultaneous retrieval, analysis and view of several thematic maps of a single geographical area. Overlaying is more or less for geography what statistics is for sociology. The exploration of hidden relationships among geo-referenced features is as creative as the correlations often explored by statisticians and professionals alike pursuing the ideals of discovering the "normal" and "hidden orders" of societies and nature. This creative potential should be kept in mind as we explore the evolution and applications of GIS, especially as new geo-referenced applications are frequently seen as true representations of both social and geographical space.

The analysis module in GIS allows single thematic maps to be queried, as well as be combined with others in order to create overlays of different spatial features. These modules include statistical, classification, optimisation, best routes and other types of algorithms that allow a myriad of operations with geo-referenced data.

In the majority of such applications it is also relatively straightforward, although not always explicit, whose purposes and interests are served by the specific overlays provided by the map: GIS was used and built by a relatively limited professional community, frequently working at the service of specific public or private interests. With the proliferation of the technology, however, it is easy to loose track of the

[1] Cf. http://www.britannica.com/EBchecked/topic/#1033394/GIS

general purpose of systems: GIS technologies are ubiquitous (Alibrandi and Baker 2008), being embedded in almost any geo-related application and artefact from weather and news reports, newspapers, internet GIS (IGIS), Internet of Things, sensors, digital globes, on-board GPS, mobile phones, games and tablet computers applications and others. As more and more of such applications go online, a number of new dimensions arise: end-to-end sharing among individual (non-professional) users, huge aggregates of information gathered from different users and places, and new software through which data mining and overlaying are interactively generated and distributed, direct integrations with sensors, etc.

### 9.1.2 The Rise of the Internet

As the Internet emerged and became part of everyday life our modes of imagining physical and social space were transformed again. Cyberspace seems to contain no boundaries; dimensions such as time, physical space, material and virtual objects acquire fluidity, dematerialise and transform time and space relationships. We are surrounded by new ICTs that are heralded as those objects, tools and concepts that will allow us to be better "connected" and more "aware" of the real world. At the same time, new barriers and boundaries are created, sometimes hidden to view in the midst of the new all-seeing visibility. Some major changes include:

First of all, maps and digital globes are becoming personalised, tailored for individual use and need through laptops, smart phones and new applications such as digital globes, Google Earth being one of the most popular. They pervade our everyday environments and activities, and the ways in which we plan and organise our geo-referenced lives. Not only do they provide far more accurate representations of concrete places, such as in Google Street View; they also radically increase the capacity to incorporate and overlay other sources of information in accordance with the needs, plans and preferences of individual persons. The quality of restaurants and neighbourhoods, travel goals, resources, business opportunities, pollution levels, art, entertainment and even friends: all of these can now be geo-located and organised according to our plans.

Second, maps become interactive: they are not made once and for all, but can be constantly changed and updated by those who use them: this goes for both "objective" and quantifiable information, but it is just as true as for experiences, fun and emotion. Whereas some of the strongest incentives for using maps and digital globes connect to the level of individual experience, planning or fun, this is having major consequences also on collective levels. In addition to more detailed visualisations, one of the most important functions of digital globes and maps is that they serve as repositories for visualising and organising large sharable amounts of information. In this way, they also become repositories for organising and restructuring whole societies. Importantly, such "sharing" of information may take on a number of different forms within different political cultures and following different world-views. We shall return to this aspect in Sect. 9.2.2. Although this article will be mainly about these first two points, we also mention the following two issues:

Third, the potential for surveillance and so-called sousveillance[2] (Mann et al. 2003) increases. Even as we may feel ourselves more free to act and to plan, we are also increasingly becoming visible to others: the more real-time Google Street View becomes, the greater the potential for everybody to *spy* on everybody; the more information we enter about our geographical and virtual movements on the Internet, the more visible we make our digital selves. Becoming voluntarily or involuntarily visible adds another layer into the representation of physical space.

Fourth, our conceptions of privacy are changing. The voluntary sharing of personal information, at a very high detail level, impacts the way we perceive public and private spheres, the way we interact and the ways in which others affect us. The level of impact is not only related to the detail of information but also to "digital memory": Once information is published online, it is very difficult to erase. In addition problems for human agency arise.

In the next section we will look further into the emergence of now popular digital globes, and the introduction of the "global" dimension in overlaying imaginations. To structure our exposition we use ideas already introduced: different kinds of overlays structure different modes of (re-)imagining the world. New and emerging imaginaries of space may be used as repositories for constructing and organising new forms of geo-referenced community or social action. We use these to describe some aspects of emerging digital and physical geographies.

## 9.2 New Configurations of Space

### *9.2.1 Earth Communities?*

Although we rarely think about it, the notion that the whole of humanity is banded together on one common planet did not come about automatically. The first images of Earth taken from space followed from the 1966 Apollo 17 moon landing. Whereas images of the earth rapidly spread through media such as television and newspapers, the actual meaning and implications of those images were wide open to different interpretations. The "space age", now possibly coming to an end with the termination of the US space programme came along with a number of new cultural expressions, such as Stanley Kubrick's science fiction movie *2001 A Space Odyssey* (1968), or David Bowie's album Space Oddity (1969). A number of earth "interpreters" were required to actively weave stories that resonated with the Earth's image, thus providing it with associations and new layers of meaning (Jasanoff 2003). In this section we focus on one such (loosely tied) community of interpreters, many of whom have come to be associated with the science of ecology or with environmentalism. Significant examples include The Club of Rome's report *The Limits to Growth* (Meadows et al. 1972), James Lovelock's *Gaia hypothesis* (Lovelock 1990), R. Carson's *Silent Spring* (1962) or R. Buckminster Fuller's metaphor *Spaceship*

[2] Sousveillance entails that participants in everyday activities are using small hand-held or wearable equipment such as smart phones, to record and monitor the activities of others.

*Earth* (Buckminster 1963). The Apollo images made possible perspectives on life on earth as bounded together within a finite ecosystem. This system was seen as fragile, vulnerable to human intervention and in need of protection. A strong claim of many early Earth "interpreters" was that the image reflected back on us, the spectators and observers, and showed us to be members of one global and interdependent community:

> I find it very important in disembarrassing ourselves of our vanity, short-sightedness, biases, and ignorance in general, in respect to universal evolution, to think in the following manner. I've often heard people say, 'I wonder what it would be like to be on board a spaceship," and the answer is very simple. What does it feel like? That's all we have ever experienced. We are all astronauts. I know you are paying attention, but I'm sure you don't immediately agree and say, "Yes, that's right, I am an astronaut." I'm sure that you don't really sense yourself to be aboard a fantastically real spaceship, our spherical Spaceship Earth. Of our little sphere you have seen only small portions. However, you have viewed more than did pre-twentieth-century man, for in his entire lifetime he saw only one-millionth of the Earth's surface. You've seen a lot more. If you are a veteran world airlines pilot you may have seen one 100th of Earth's surface. But even that is sum totally not enough to see and feel Earth to be a sphere unless, unbeknownst to me, one of you happens to be a Cape Kennedy capsuler (Buckminster Fuller 1963, ch. 3).

The images of Earth that arrived with the space age created global images that could traverse national and cultural boundaries; they gave birth to a new "global" discourse about Earth, its vulnerability and limitations. The lasting outcomes of these contributions can now be seen in the importance given to concepts of sustainability, precautionary policies, intergenerational responsibility and global biodiversity policies. But one thing is to project such shared fate and global community; quite another is to call into being such community and to effectively act together. Commentators such as Buckminster Fuller remained restricted to an intellectual elite with *its* ideological commitments (in many cases coming out of leftist movements, starting to transform them in environmentalist, "green" directions), and heavily reliant upon certain forms of scientific discourse, such as systems theory and economics. As clearly seen from the failings of the Copenhagen 2009 climate meetings: it is one thing to make a diagnosis of the suffering Earth; quite another is to arrive upon coordinated and effective action to halt threatening developments such as deforestation and draught, global warming and resource depletion. Interestingly, Buckminster Fuller had great faith in computers and their power to free humanity from its negative predicament. But even he could not have predicted the many and diverse forms in which the globe is now depicted, represented and reproduced through a number of applications on laptops, handheld devices, phones and new media.

### 9.2.2 Digital Globes

The first pictures from space were soon followed by satellite systems' imagery allowing for constant visualisation of the Earth terrains. Before this, aerial photography had been too expensive to be "accessible" in democratic ways. Satellite imagery have democratised the access to Earth imagery, since digital maps became cheaper and increasing computer power allowed competent map processing by non-professional users. Indeed, these have now become directly accessible in personal computers. Interesting new possibilities emerge as soon as the Earth globe can be downloaded to laptops and smartphones, to be manipulated with, exchanged and used for a great number of purposes: a next generation of "scientific and digital interpreters" emerges in all areas where geo-location is relevant. Google Earth, one such system, offers services that render scientific and ecological data directly accessible, combined with a strong experiential base *and* more realism:

> Google Earth allows you to travel the world through a virtual globe and view satellite imagery, maps, terrain, 3D buildings, and much more. With Google Earth's rich, geographical content, you are able to experience a more realistic view of the world. You can fly to your favorite place, search for businesses and even navigate through directions. It's all up to you! (Google 2011)

Such developments created the conditions for reproducing and re-imagining the lines from the previous section. The Earth as limited and vulnerable also finds its way into new, personalised geo-referenced applications. As described by one participant in a online discussion forum on ethical and social aspects of GIS,[3] when combined these may have powerful effects:

> [C]ontemporary global "live" maps work on our minds along with the now shared opinion that the globe is not well. […] What we are seeing is not just an outstanding planet. We are looking at ourselves. What we see is a resultant of what nature and we jointly do. "What on earth are we doing?" presses itself upon all of us. If we want a closer look somewhere, we can, by way of Google Earth zoom in on the neighbours CO2 emitting lawn-mower, and then back out again to the earth floating in space. If we miss the live view of the details, chances are there will be a webcam to observe for us, the cam being a natural extension of the map. (perarian, Technolife forum participant (de Sousa Pedrosa and Vanderlinden 2011))

Al Gore, the US politician also known for the movie *An Inconvenient Truth*, was one of the main promoters as Google Earth developed its own climate change application. In Google Earth Climate one may focus on one's own home, or any region of the globe, and "see" climate projections for that area (based on the Intergovernmental Panel for Climate Change (IPCC) reports). In his book called *Our Choice* (2009) Al Gore takes such themes and applications one step further in the creation of an interactive book, for publication on devices such as iPads and

[3] The project was called Technolife and was funded by the European Commission. The aim of the project was to create debate on ethical issues relating to uses of new technologies, such as digital globes. The project included an open forum to which people that are concerned by the development of digital maps were invited to discuss the above mentioned issues. More information at: www.technolife.no

iPhones. *Our Choice* uses interactive pictures and graphs, as well as contributions from scientists and filmmakers, to provide an overview of solutions to the climate change problems. Special focus is on different forms of renewable energies. In the promotion of the book Gore claims that *"Our Choice gathers all of the most effective solutions that will solve this crisis"* (Gore 2009). In comparison to the previous discourse on the vulnerability and finitude of the Earth, Gore's approach is considerably strengthened, by the underlying conviction provided by the IPCC reports, but also by the instrumental use of new media as a way of influencing public opinion. The realism and emotional content provided by digital globes and maps not only work to convince through argument, as did Buckminster Fuller. It also works through the strong persuasive force of images and computer-mediated interactive applications.

Even though one may sympathise with the diagnosis put forward by Gore, and share some of his wider goals to produce global change, it is also clear that he strongly uses this power, of visualisation and interactivity, for purposes defined by him. The interactivity of *Our Choice* is as many applications provided through Apple a one way-street: There is no way in which users can change or challenge the content of the book; upgrades are exclusively done by the publisher. In this sense, *Our Choice* continues a strong tendency from Buckminster Fuller: overlaying and interpretation of the fundamental facts is provided by scientific and political elites. Although intentions may be good, the following problems also follow: what about those who do not share the diagnosis or solutions provided by Gore's book? And how should one relate to the fact that he, and many of his collaborators, have significant economic interests in the kinds of solutions suggested, such as the rapidly emerging renewables industry? This points to a general problem with changing our societies towards sustainability: there is a need to consider the cultural specificities of the attitudes of the citizens and communities whose behaviour is to be changed by elite initiatives. Do they identify with the visions provided by leading "Earth politicians" such as Gore? There is also a need for considering the democratic aspects involved: who gets to define the problem, and who gets to define its solutions? These questions are not external to digital devices and applications used for sustainability purposes, but rather intrinsic to them. In the next section we describe a different approach to issues of interpretation, overlaying and interactivity.

### 9.2.3 Lay Cartographers: Sensoring and Layering the Earth

Topographic and satellite imagery become more interesting through bottom-up initiatives and social movements through which people worldwide contribute to the creation of new geographies. Such initiatives may take different shapes according to technologies and types of overlaying, but importantly also to how they reconfigure previous relations between the private and the public. Communities of *volunteer virtual cartographers* are actually making maps, which often have great impact in society. Their maps are several times used as base layers by privately own applications (such as Google maps), organisations and governments. Due to their capability for promptly creating new information (as a consequence of the many distributed contributors) they have a potential hardly comparable to standard institutionalised methods. Such initiatives can make a real difference, for instance, in disaster situations.

One prominent example is the *Ushahidi*, an open web platform initially developed by activists and programmers to map reports of violence in Kenya after the post-election fallout at the beginning of 2008 (http://www.ushahidi.com). Ushahidi means "testimony" in Swahili, and the name refers to the ways in which the platform enabled citizens and NGOs to report incidences of violence outside of established state and media channels:

> The Ushahidi engine is there for "everyday" people to let the world know what is happening in their area during a crisis, emergency or other situation. Bringing awareness, linking those in need to those who can assist, and providing the framework for better visualization of information graphically. (Ushahidi 2008a)

The original platform allows cell phone users to tell their stories and geo-locates them by placing them on a Google map. Only stories that could be verified by additional sources were posted. Testimonies and messages were grouped into distinct categories according to their content: "Riots", "Deaths", "Property loss", "Government forces", "Civilians", "Looting", "Rape", "Peace efforts" and "Internally displaced people" (Ushahidi 2008b). The following quote is an example of one report categorised as simultaneously "Riots", "Deaths" and "Government forces":

> Just before celebrating the first month of tranquility in Mathare, The Kenya Police has provoked the volatile community to violence ending in the death of an infant, slashing of people with machetes and burning of a Matatu commuter bus to ashes in Mathare North. Once again as is synonymous with other opposition strongholds, Police violently arrested residents of Mathare North indiscriminately. Men women and Children were not spared in the raid that was characterized by rapid gun shots equaling the "commando" movie by Arnold Swaz to drive away residents from watching the on goings. We decry the barbaric, insensitivity and the careless handling of a pregnant woman who latter in the day gave birth at Kasarani Police Station. As is common with the dreaded fugitives, the woman was ferried to the hospital as if Osama Bin Laden had just been arrested by Mr. Bush. With the sirens all over the air, and potholes of Nairobi, the baby was thrown up and down in a rough ride to death (Kennedy 2008).

Compared to a situation in which such incidents are allowed to happen outside the scrutiny of national and international public attention, it is clear that the Ushahidi platform is a strong tool for empowerment. The designers of the platform, being activists rather than business or government people, make available and facilitate a certain service to connect previously unconnected individuals and sources of information. In so doing they provide channels for the creation and communication of alternative actions and so also open up for the imagination of alternative futures. From a scientific point of view, this may render data inaccurate and of little relevance to experts. However, for a great number of other purposes, such as social mobilisation, it is exactly experience, emotion and concern for one's environment that attract people to geo-referenced applications. Violeta, another Technolife forum participant, argues:

> (…) the social production of maps, normally online, is becoming such a powerful tool to enhance the comprehension of territories (including space and people), social processes, expressing denounces, building participatory proposals, and thus taking part on governance, for not GIS experts. In this sense, accuracy is not so important since the geo-referenced data is normally quite simple, but it strength is its democratic nature (from de Sousa Pedrosa and Vanderlinden 2011)

Later on, the Ushahidi has also been used in other situations of crisis, for instance during the Haiti humanitarian disaster. Importantly, users of the software are free to change it according to their own needs and purposes, and to re-publish new versions. In contradistinction to the above approach of Al Gore, activist uses of GIS tend towards Open Source solutions, and so seem to be attending more also to the democratic aspects, not simply of the actions facilitated by the technology, but also of technological innovation itself.

### 9.2.4 Living in a Privatised World?

The above mentioned examples differ, both in terms of technologies used and in terms of the kinds of public/private relations *instigated* by them. However, they are typically regarded as committed to the promotion of the care and maintenance of *common spaces* and interests of a public nature. That, however, is not a necessary consequence of new GIS and digital globes applications: In a number of new GIS-applications, the line between promoting and weakening public spaces is blurred. Steve Graham (2005) has introduced the concept of "software-sorted geographies" to describe how geo-demographic information systems are used to characterise neighbourhoods in terms of crime rates, accessibility to services, environment, pollution, and school performance. Collective spaces are reconfigured through online applications giving easy access to seemingly objective and neutral information about living spaces and communities. These tendencies may be seen as especially prominent in the Anglo-Saxon world, but also in a number of Latin American countries where socio-economic differences are high. One example from the UK is Upmystreet.com, which gives information on a number of lifestyle related factors and issues, such as gyms, schools and shopping, as well as the people living in specific neighbourhoods. Before buying a house one may investigate the community and neighbourhoods across multiple layers of meaning and values. Another interesting example in that respect is offered by the webpage MapTube (http://www.maptube.org/), which maps "anti-social behaviour". Categories range from "Drunken Youths" to "Noisy neighbours", to "Boy Racers" and "Great Community and No Problems".

More and more choices about where to live are nowadays made through online services. Insofar as these sites are used as reliable sources for choice, the online world defines the quality of the "real" world, the geographical and social environments in which people live. So-called "gated communities" provide secure and comfortable living spaces for the affluent, whereas keeping out un-wanted elements, but also effectively the rest of society. These types of spaces are strongly reflected and promoted through new digital environments. Here is a quote taken from the Forbes Magazine article "The most expensive gated communities in America". The post is about the rating (in "spots") of those communities:

> ...I live in the Gaurd Gated community of Mulholland Estates in Beverly Hills, CA, and it sounds as if it deserves at least a 5–10 spot. It is 5 min down Mulholland from #2 Bevery Park (which I believe is the true #1) it has about 75 homes, 24-armed security, its in the hills w/ city views, and homes have been selling for no less than 5,000,000 up until about 15,000,000. Robbie Williams (I think he's a British celeb right?) lives down the street, as well as Christina Aguilera, Shaq recently moved, Tom Arnold lives here, as well as Loni Anderson, and Judge Mathis. But, celebs or not the community is pretty cool...you should look it up. (Forbes 2006).

Graham uses the term "un-bundling" in order to describe how such applications may work to decompose communities and public spaces: online representations of neighbourhoods and communities reconstruct public and common goods as consumer goods and matters for individual preference. The "community" is defined by money, status and security is an important value. In the above case one single person provides the concrete overlay. But, of course, that expression is made possible by the service provider. Through such online platforms choices and selections are promoted in which space is reconfigured according to the imperatives of marketing and consumerism.

## 9.3 Discussion and Concluding Remarks

More than a decade ago, Batty (1997) suggested that entire new geographies (virtual geographies) were being constituted within the computer with little resemblance with the real geographies and moreover that the real geographies were being changed through virtual communication. He attributed these changes to the changing notions of space and place with computers and communication. Finkelberg (2007) introduces the notion of *neo-geographer*, a map user that makes use of technology previously restricted to professional domains. An amateur map user, whose agency, knowledge, and input into the creation of the map is certainly different from a professional one, does not require representations of space that stress accuracy and precision. Unlike the urban planner or erosion specialist, the neo-geographer use of digital maps reflects lived experience. The *neo-geographer* is all of us who use digital globes or any geo-referenced application. Freire and Villar Onrubia (2010) suggest that the emerging new-geography and cyber-cartography practices are being the "platforms" through which citizenry are being motivated to participate and culturally (re)appropriate the public urban space, as well as the stimulus for construction of urban imaginaries.

Overlaying overcomes old physical boundaries, creating other boundaries that have little to do with physical space as we are accustomed to think about it, i.e. the space that is published and official and describes a specific institutional order that few have established. Overlaying generates and aims at establishing and probing other rearrangements of geography, politics, communities, etc. and creates newer visible or concealed orders. The overlaying activities have been changing the ways in which we conceive of space and of its properties. This is not only about newer actors taming the concept and properties of space, but also about the ways emergent technologies are enacting those processes and their appropriation – a clear example of how technologies and social meanings co-evolve. In accordance with this we have argued that different modes of imagining and representing the world are literally constructed through different overlaying of data, embedding in geographical systems the many and wider cultural and socio-political settings. Overlaying is a mental and technological process through which we imagine the world and our place in it. It is about choices of what can be seen and considered, similar to how other technologies such as statistics and probabilities are used to create collective representations (ultimately to establishing the "normal"). However digital globes brings overlaying to a completely different level of aggregation and potentiality, with impacts still largely unknown. Almost everything that we do is now geo-referenced, even the photos we store in the Cloud through the myriad Apps available. This raises a number of important questions that could not be touched upon here, such as: What happen to agency, anonymity, privacy and reputation when people and all their activities become geo-referenced and available to scrutiny by "everybody"? For who and for what purposes are data collected, stored and processed? What kinds of mechanisms are put in place to guarantee quality, security, privacy, dual use, etc.? What happens to quality when we start talking about "big data"? Or are these no longer issues? Can one opt-out from the new digital layers, become invisible?

Two issues of main importance have been dealt with in this paper. First, we have shown how space was disciplined, and how, at the inception of the modern age, this was intimately connected to a conception (and indeed a kind of hope) of space as embodying objective and neutral properties. At the same time this provided the means for a far more powerful political enterprise, map-making, than perhaps foreseen. Both of these aspects of space are today reproduced and reconfigured in ways too numerous to be dealt with in their entirety (one central example is the self-presentation of Google Earth as provider of more realism and fun at the same time). It is clear that space was never a single sphere or thing, and that as we go digital, ubiquitous and different spaces are promoted and produced. Second, we highlighted (albeit generally) some ways in which space, as a mediator of human interests and communities, takes on new, self-generating traits in the digital age. As we continue to make collective choices for what space might become in the future, what kind of social imaginaries, interests and self-generating social and technical mechanisms are nurtured and allowed to grow? Although also dependent on individual user choice and interaction, these are fundamentally collective dynamics and choices. Strong imaginaries of space as a private commodity will promote private space and financial power over communal projects and spaces.[4] On the other hand, we also observe forces driving in other directions, as when local communities and NGOs contribute to environmental projects, or when citizens are given new mechanisms for holding authorities accountable.

Important aspects of the problem complex relate to impartial and incomplete knowledge of the enormous complexity, the many layers, of spatial information. These issues cannot be completely disconnected from issues of social justice, power and access to information: even as we think we work for the "good", we may be generating information highly useful to people and organisations with other agendas and interests. This is what the ICT ethicist Luciano Floridi (2006) has termed the Tragedy of the Good Will.

On the background of reflections such as these, we would end with the following question: even though we cannot maintain space as a unitary concept in science, perhaps are we well advised to keep it as a valuable and flexible concept when we think about political, democratic, ethical and environmental issues?

**Acknowledgements** This paper was made possible by the research opportunities emerging from the FP7-funded project TECHNOLIFE (project number 230381).

## References

Alibrandi, Marsha, and Thomas Baker. 2008. A social history of GIS in education, 1985–2007. In *Digital geography: Geospatial technologies in the social studies classroom*, ed. Andrew J. Milson and Marsha Alibrandi. Charlotte: Information Age Publishing Inc.

Batty, Michael. 1997. Virtual geography. *Futures* 29(4/5): 337–352.

Britannica Encyclopedia. 2011a. *Space*. http://www.britannica.com/EBchecked/topic/557313/space. Accessed 3 Aug 2011.

Britannica Encyclopedia. 2011b. *Carthography*. Britannica Encyclopedia. http://www.britannica.com/EBchecked/topic/97492/cartography. Accessed 3 Aug 2011.

Buckminster Fuller, Richard. 1963. *Operating manual for spaceship earth*. New York: E.P. Dutton & Co, Ch. 13.

Carson, Rachel. 1962. *Silent spring*. Boston: Houghton Mifflin.

de Sousa Pedrosa, Tiago, and Jean-Paul Vanderlinden. 2011. *Technolife deliverable D3.5, D3.6, D3.7: Documentation on ICT, GIS and body deliberation*. Technolife consortium, Versailles, annex XII.

Finkelberg, Amanda. 2007. *Space, place and database: Layers of digital cartography*. MSc thesis, Massachusetts Institute of Technology, Massachusetts.

Floridi, Luciano. 2006. Information technologies and the tragedy of the good will. *Ethics and Information Technology* 8(4): 253–262.

[4] Another example that could not be dealt with in this article: The *militarisation of space* feeds on imaginaries of "the others", as when the US Government maintains some degree of control over mapping and real time geo-referencing systems, and so over navigation with the global GPS system. This happens at the same time as European and Chinese authorities seek ways to break the monopoly (e.g. the European Galileo Project).

Forbes. 2006. *Comment to article*. The most expensive gated communities in America. http://www.forbes.com/2006/10/09/gated-community-homes-life-re-cx_lm_1010gates.html. Accessed 20 July 2011.

Freire, Juan, and Villar Onrubia, Daniel. 2010. *Prácticas cartográficas cotidianas en la cultura digital*. Razón y palabra. in http://dialnet.unirioja.es/servlet/articulo?codigo=3266254. Accessed 3 Aug 2011.

Google. 2011. *About Google Earth*. Google Earth. http://earth.google.com/support/bin/answer.py?hl=en&answer=176145. Accessed 3 Aug 2011.

Gore, Al. 2009. *Our choice*. Push Pop Press. http://pushpoppress.com/ourchoice/ Accessed 3 Aug 2011.

Graham, Stephen. 2005. Software sorted geographies. *Progress in Human Geography* 29(3): 562–580.

Jasanoff, Sheila. 2003. Image and imagination: The formation of global environmental consciousness. In *Changing the atmosphere: Expert knowledge and environmental governance*, ed. Paul Edwards and Clark Miller. Cambridge, MA: MIT Press.

Kennedy, Owino. 2008. *Arbitrary arrests sparks protests*. Ushahidi. http://legacy.ushahidi.com/incident.asp?id=218. Accessed 3 Aug 2011.

Lovelock, James. 1990. Hands up for the Gaia hypothesis. *Nature* 344(6262): 100–102.

Meadows, Donnella, Meadows, Dennis L., Randers, Jorgen and Behrens III, William W. 1972. *Limits to growth*. New York: Universe Books.

Robinson, Arthur H. 1986. Which map is best? American congress on surveying and mapping, American Cartographic Association's Committee on Map Projections (ACACMP) Falls Church.

Ushahidi. 2008a. *Ushahidi 1-Pager*. Ushahidi. http://www.ushahidi.com/-/docs/Ushahidi_1-Pager.pdf. Accessed 3 Aug 2011.

Ushahidi. 2008b. *The original Ushahidi platform*. Ushahidi. http://legacy.ushahidi.com/. Accessed 3 Aug 2011.

# Chapter 10
# Beyond Professional Ethics: GIS, Codes of Ethics, and Emerging Challenges

**Fanny Verrax**

**Abstract** This chapter questions the adequateness of GIS (Geographic Information Systems) professional ethics by analyzing the URISA's ("The Urban and Regional Information Systems Association": the broadest association for GIS Professionals) Code of Ethics. It starts by a preliminary mapping of ethical issues raised by GIS. Its intent is to go beyond the traditional PAPA issues (Privacy, Accuracy, Property, Access) by taking into account issues such as space as a construed object, individual identity and the issues of scope and scale. After exploring various perspectives on professional codes of ethics, it considers how the field of GIS professional ethics have dealt with these issues and suggests a dichotomy between academic and professional ethics. It finally suggests to rethink the expert-lay people interplay when discussing GIS ethical issues.

**Keywords** Philosophy of Technology • Technoethics • Geographic Information Systems • Professional codes of ethics

## 10.1 Introduction

Along with nanotechnology and biotechnology, geotechnology, including Geographic Information Systems (GIS), has been identified by the US Department of Labor as one of the three most important emerging and evolving fields (Gewin 2004). Geographic Information Systems are a tool now used by all industrialized countries and their governments in order to represent different ranges of spatial data. What makes GIS essentially different from other ICTs (Information and Communication Technologies) is that it integrates traditional database with spatially referenced data in one organized system. "Its inherent power lies in its flexibility of manipulating spatially referenced information about the data and displaying it as snapshots of real life phenomenon" (Haque 2003).

F. Verrax (✉)
National des Sciences Appliquées (INSA) de Lyon, Villeurbanne, France
e-mail: fanny.verrax@gmail.com



A. Delgado (ed.), *Technoscience and Citizenship: Ethics and Governance in the Digital Society*, The International Library of Ethics, Law and Technology 17,
DOI 10.1007/978-3-319-32414-2_10

Most of the literature about GIS code of ethics focuses on improper use of GIS that is due to the lack of competence or limited knowledge about GIS technology. However, according to (Haque 2003), “a large part of GIS use has to do with ethical issues that transcend technique, and highlight the need for a broader understanding of the role of GIS in a democratic society.” It insists that the argument is not whether information and its related technology is intrinsically good or bad, but that “the primary concern would be for citizens to ensure that government officials, who have access and control the information, hold it in trust.” (Haque 2003).

How have ethical issues arising from the use and development of GIS have been considered and identified in the past? How do professional ethics address these issues and are current codes of ethics of GIS professionals sufficient? Are there other issues or challenges that should be addressed, and how?

## 10.2 Preliminary Mapping of GIS Ethical Issues

### *10.2.1 Some Lessons Taught by Information Ethics: The PAPA Issues*

As any new technological device, Geographic Information Systems rise ethical challenges: “how to use them in order for them to benefit society as a whole and not only private interests? Could they harm individuals or collectives and how to avoid that? Are there any uncertainties related to its quick and massive spread and what are they?” are some of the questions which could apply to most technological break-through, and among them GIS.

Three related conceptions of GIS are usually accepted, based on space, rationality or information (Curry 1995). Thus, issues related to information ethics may be relevant when considering GIS. Furthermore, they have been better investigated than other issues since information ethics are a quite well-developed field. In its pioneer paper “Four Ethical Issues of the Information Age”, (Mason 1986) identifies four major ethical issues which are to be thought about in any activity involving information, and which are summarized by the acronym PAPA: Privacy, Accuracy, Property, Accessibility.

#### 10.2.1.1 Privacy

This issue is definitely the most important one, or at least the one that has received most attention both from the medias and the general public. For instance, in 2010, 250,000 inhabitants of Berlin opposed the introduction of Google Street View in their city spaces and asked Google to block the images of their houses. “These citizens experienced Google Street View as a technology that allows them to watch and to be watched.” (Sune Jepsen et al. Chap. 2, in this volume). Another relevant example can be found on the web: a “Yahoo Question” from 2009, asking: “List 5 ethical issues that arise from Google Earth?” definitely puts privacy as the leading issue:

(From http://answers.yahoo.com/question/index?qid=20081220224110AAy7rX7)

Indeed, two forces in GIS such as Google Earth are said to threaten privacy, the growth of information technology, and the increased value of information in decision making. Usually, the danger arises mostly from merging current databases into one centralized data bank, what (Mason 1986) calls "the threat of exposure by minute description".

Some more recent GIS, combined with Web technologies, may reinforce even more this public concern over privacy: "Specific GIS applications with strong ethical considerations include geodemographics, which has the ability to help invade personal privacy. With the advent of ever more interlinked Web portals and publicly available advanced search engines, the combination of GIS and Web technologies has opened up new routes by which privacy can be attacked for legitimate (law enforcement), illegitimate (cyberstalkers) and disputed reasons (unwanted direct marketing)." (Longhorn 2004). Another danger of centralized database is the repercussion and amplification of possible mistakes, which leads to the issue of accuracy.

#### 10.2.1.2 Accuracy

As stressed by (Mason 1986), "today we are producing so much information about so many people and their activities that our exposure to problems of inaccuracy is enormous." Another challenge about accuracy can also be found in the format of the information. Indeed, there are numerous examples of information available but in the wrong format, with incorrect resolution, or being incomplete.

#### 10.2.1.3 Property

Two main questions arise from the issue of property: who owns the information displayed in GIS, and how are contributors of a particular knowledge to be compensated? About the first question, two models have been competing forever: privative access versus open source. (Mason 1986) makes a comparison between the traditional pasture as a "common" and the bandwidth as a common, which leads to the questions: "How will the limited resource of bandwidth be allocated? Who will have access?". About the second question, (Mason 1986) focuses on expert systems to ask: "Is this exchange of property [between the human experts who provide the information and those who own the hardware and software] warranted?". He also proposes a parallel with the steam energy industrial society and Jacquard's weaving looms, in which weavers suffered unemployment and degradation.[1] Indeed, the looms used the technical expertise of weavers and made it at the same time useless, since much less weavers were needed.

#### 10.2.1.4 Access

According to (Mason 1986), in an Information Society, the main ethical challenges in order to ensure equity in access to GIS are much broader than just the design or property of GIS: it concerns the society as a whole, regarding how the citizens have access to education, and what kind of technical and cultural facilities they are provided.

### *10.2.2 PAPA as an Ethical Dilemma*

Another remark ought to be made about the links between those four issues (Privacy, Accuracy, Property, Access): although they seem to be all as desirable, there is a challenge not to be ignored about their compatibility. The most striking case is between the issue of privacy and the issue of access: if the privacy of personal data is fully respected, then the access to potentially useful data is very limited. On the contrary, if access is to be encouraged, privacy will eventually not be fully respected.

It is an eternal legal and ethical dilemma between citizen right to access information, reaffirmed by the UNESCO in 2000 ("Governments need to balance their strategy between preserving the integrity of intellectual property rights (IPRs) and the need for broad access to information and knowledge.") and protection of personal

[1] This weaving loom is a mechanical loom, invented by Joseph Marie Jacquard in 1801, that simplifies the process of manufacturing textiles with complex patterns. The loom is controlled by punched cards with punched holes and is therefore considered an important step in the history of computing hardware, but led to major social protests ("révolte des Canuts" in Lyon, France) and Jacquard himself regretted his whole life the social consequences of his invention.

privacy – protected for instance by the Universal Declaration of Human Rights, Article 12 ("No one shall be subjected to arbitrary interference with his privacy, family, home or correspondence, nor to attacks upon his honour and reputation. Everyone has the right to the protection of the law against such interference or attacks.".) In more technical terms, those two issues remind us of the concept of "compossibility", as defined by Leibnitz: two desires, or two states of the world, are said to be compossible if they can exist at the same time, which is not the case if we consider two ideal states of privacy and access, which could therefore be considered "incomparible" issues to a certain extent: it is not possible to have a GIS displaying full geospatial data, with no limitations of access whatsoever, and still not threatening anyone's feeling or desire for privacy.

Subsequently, we can identify, within the PAPA scheme, an ethical dilemma, which is traditionally defined as the conflict between two or more moral imperatives. A typical way of addressing such a dilemma in classical approaches to ethics would be by either focusing on the practical consequences of the action that is considered, or focusing on the action in itself, assessing whether it is right or wrong. For instance the fact that Google Street View takes pictures which allow identifying individual houses, by threatening privacy but improving access to geographical information, is definitely an ethical dilemma that can be framed in both approaches. One can say that capturing and storing these kind of images is morally wrong, no matter what practical consequences will occur, or one can focus on the practical consequences it may have, both positive and negative: has it increased burglaries in the neighborhood? Is it easier to find the place you want to go now? Have the real estate prices dropped or raised or stayed the same? etc. Of course one difficulty of this latter approach is that very often in the case of GIS, it is not the consequences which occur *now* that are feared, but the ones which might happen in the future, and are therefore very delicate to identify.

Therefore, I disagree with (Böhm and Pfister 2000) when they distinguish two judgmental aspects of risk evaluation: evaluation of consequences and ethical considerations. Opposing taking into account consequences and ethical considerations is not a fruitful way of solving an ethical dilemma. Instead, considering consequences is a full part of ethical judgment. Such an analysis would avoid a limited idea of risk (as probable consequences that can be quantified) and would rather focus on uncertainties, unknowns and values. In this light, ethical dilemmas such as that presented above cannot be solved by a technical solution: it requires a debate between holders of different but explicit values systems.

### 10.2.3 Technical Logic Versus Social Logic

This tension between two logics is not specific to GIS, nor to Information Ethics. It arises in many fields and has been discussed about many technological developments. From the point of view of GIS as sciences (GIS standing then for Geographic Information Sciences instead of Systems), they are made up of a dynamic interplay of both a technically oriented and top-down logic (i.e. from the experts to lay-people) and a bottom-up logic (i.e. from citizens to GIS professionals). Those two logics can be understood as coming under different dynamics requiring different ethical frames.

In philosophy of technology, two lines of arguments have been developed about technical systems, which are radically opposed. The first one could be seen as a deterministic point of view. It affirms the autonomy of technology, which has been good to some point, but could now become dangerous for humans, especially if we lose control on it. It is basically the thesis developed by the French philosopher of technology Jacques Ellul and his many disciples. On the other side we find the thesis of the neutrality of technology, often developed by former engineers or natural scientists and who strongly disapprove the first line of arguments.[2] Basically, it says that any technical system is nor good nor bad, but just a mean to achieve a certain end, which is determined by humans, and that moral judgments or ethical issues should focus on this end and not on the means which are just means and by definition neutral. However, are these two approaches opposed or complementary? Every technology has a set of values inscribed in its design, whether they become realized will depend on the ways in which the technology issued and regulated to a large degree.

In the case of GIS, many are those who conceive to see it as an autonomous system, at different levels though. According to (Chrisman 2006), the claim for "technology autonomy" is based on two major tenets: first, technology is seen as engaged in a "March of Progress", which leads from less to more advanced systems; second, technology is an imperative to which social institutions and people must adapt. Michael R. Curry, a geographer interested in cultural geography and the history of geographic ideas affirmed therefore that "in the case of GIS this leads to a kind of fatalism, where discussions about the desirability of the development and use of the systems are seen as frivolous attempts to push back the tide." (Curry 1995). This reminds us of the *Gabor* rule ("loi de Gabor") stating that everything that is possible will be necessarily realized, or that every possible combination between available technologies will be exhaustively attempted. According to M. R. Curry, this belief can lead to heteronomous behavior – which is, non-autonomous, in the sense that humans, and among them decision-makers, somehow give up their power to make decisions, convinced that it is pointless anyway. This belief also leads to the increasing complexity of technical systems, which has two mains effects. First, this means that technological systems have become "black boxes", from which it is impossible to predict results, hence teleogical value systems, which is systems that are aiming at a specific predefined goal, become impossible to apply (Winner 1986). Second, that complexity leads to an increased division of labor, of which two primary expressions are specialization and expertise.

M. R. Curry's conclusions were that GIS development would involve necessarily ethical inconsistencies. These inconsistencies would not be merely contingent features of current practice but necessary features of the system. Whether or not he is

[2] Cf. for instance the German philosopher Friedrich Rapp.

right, it seems therefore of paramount importance to develop specific frames in which ethical issues raised by GIS development and use could be addressed properly.

### *10.2.4 Going Beyond PAPA: The Specificity of Ethical Issues Associated with GIS*

If the issues of privacy, accuracy, property and access appear to be quite relevant for GIS Ethics, as well as the one opposing top-down and bottom-up approaches, it also seems that GIS display a number of characteristics which raise further ethical issues, which may not be able to be addressed merely by a general frame of ICT ethics. This appeals to what some call "the uniqueness debate": do geographic information systems raise ethical questions that are specific or which can be reduced to old, well-known issues in traditional ethics? The uniqueness debate has been quite vivid during the early development of information ethics, until the majority of both communities, ethicists and computer scientists answer in the affirmative. In the case of GIS, the main specificity of the ethical issues it raise, although not well recognized at the present moment, consist in three aspects.

#### 10.2.4.1 Space as a Constructed Object

From the point of view of the object itself, the space as a treated object allows representing, after manipulation of a great amount of data, different phenomena as they are inserted in a physical space of personal or social action. And yet these representations are heavily embedded within the cultural and paradigmatic frames which are dominating in the community of GIS practitioners and potentially, they can be imposed on human communities living in the treated spaces. Even worse, GIS practitioners consider massively their object, space, as being "objective" and independent from any subjective belief. Here, collaboration with ethicists and philosophers of science would prove useful in the realization that there is no such thing as an "objective object", and that paradigms in the history of science are indeed very often influenced by cultural and social factors. Considering space as a constructed object would make both GIS practitioners and GIS users aware of the political and social content necessarily present in any GIS. Borders between countries are a typical example. Further features proposed by Google Earth make it even more obvious: what does it mean when a particular place is covered by Panoramio pictures while vast spaces seem completely empty and somehow uninteresting?

In this light, the issue of accuracy appears as having important political dimensions. When a country declares its independence, or when a new territory is created, the fact that the new name is shown or not in a map is of paramount importance. And even when the new state of facts is not contested by anyone, showing it on a map can take some time. For instance, the territory of Nunavut in Canada, with a mostly Inuit population, was created on April, 1st 1999, from a separation from the Northwest Territories, but only appeared on Google Earth in 2006. And still today, even though is it the largest federal territory of Canada, its name requires the biggest zoom in order to appear.

#### 10.2.4.2 Individual Identity and Living Space

At last, Geographic Information Sciences process data which affect individual identity, in its spatial dimension, which is a dynamic defining element of it. Affecting individuals' space may always transform them. Besides, the existence of huge database connecting personal and spatial information cancels out what (Curry 1995) called "the possibility of redemption". Indeed, our individual identity can be defined as selectively making public certain things about ourselves. One important feature of this process involves memory: we rely on the fact that there are things about us that others will forget, and that there is "the possibility of redemption". A highly organized surveillance system denies that possibility. Paradoxically, it goes with a "license to forget": we need to remember less and less as we have access to more and more information, as pointed out by (Winner 1977). Furthermore, digital information tends to update itself, the past becoming erased as in Orwell's science-fiction book *1984*. In the case of GIS, this can be illustrated by the growing habit to rely on updated maps available continuously on our different technical artifacts (personal computers, mobile phones, GPS, etc.). This allows us not to remember anything about the geographical space we are living or travelling in. It is therefore essential that conceptual progress in the ethics of living space shall be integrated in the ethical reflection on geographical sciences.

#### 10.2.4.3 The Issues of Scope and Scale: From Technical to Human Issues

GIS are currently facing two issues that are presented as being only technical but arise in fact mostly from our human and political conception of how science and technology should be ruled. The first issue is the one of scope: "simply there is too much data covering too many topics for GIS professionals to have enough bodies and subject matter expertise to cover it all." (Gorman 2011). The second issue is the one of scale: "the ability of GIS professionals and their technology architectures to deal with the volume and speed of data creation." (Gorman 2011) "How could these issues *not* be technical?" are you probably wondering. Isn't it just an issue of hiring enough GIS professionals, having enough computers, enough softwares or whatever it takes to make GIS run? Well, as stressed by (Gorman 2011), "The scope problem is really driven by the GIS dogma of authoritative data" in the sense that in most of GIS, data either needs to be created by a GIS professional or validated by them. A more participative, bottom-up approach could then contribute solving the issue of scope. But this is again a matter of choice: do we, regular citizens and users of GIS, prefer to rely on a technology that is created by experts for our own heteronomous use (remember the difference, heteronomous as opposed to autonomous), or do we want to try to understand, collectively, how this works, and be a part of it? And how do GIS professional codes of ethics allow and frame that choice?

## 10.3 Professional Codes of Ethics: Issues and Challenges

### *10.3.1 Where Do Professional Ethics Come From?*

In a broad meaning, professional ethics have always existed, as they arise from an inevitable tension "between the professions' pursuit of autonomy and the public's demand for accountability" (Frankel 1989, p. 109). Already in the 1950s British philosopher Bertrand Russell talked of professional ethics, as being somehow separated from the ethics of the ordinary man.

> Each profession comes to have its own ethical code, in part different from that of ordinary citizens, and in the main more positive. Doctors are bound by the Hippocaratic oath, soldiers by the laws of military discipline, priests by a number of rules from which other men are exempt. Kings must marry as the interests of the State direct, and not according to the promptings of their own inclinations. The positive duties belonging to each profession are in part prescribed by law, in part enforced by the opinion of the profession or of the general public (Russell 1954, p. 28).

Indeed the examples he gives concern a minority following a vocation, be it medicine, the Military, the Church or – the least vocational of all maybe – being part of a Royal Family (had Sir Russell not be British though he could probably have written the same about political leaders in general). Years later, (Chadwick 1998) writes:

> Professional ethics is concerned with the values appropriate to certain kinds of occupational activity, such as medicine and law, which have been defined traditionally in terms of a body of knowledge and an ideal of service to the community; and in which individual professionals have a high degree of autonomy in their practice.

As it appears from these two quotations, medicine is the paragon of professional ethics, which is easily explained by its vocational aspect, and medical practitioners can rely on a large body of ethical guidelines that have reach strong consensus over the years. In a more restricted and technical meaning however, professional ethics are a phenomenon that arose mostly in the second half of the twentieth century – an academic thesis from 1965 identifies the birth of codes of ethics outside the fields of law and medicine to three decades earlier, that is the 1930s for the very beginnings then (Brown 1965). How did these codes emerge and what are their functions?

### 10.3.2 Modern Codes of Ethics

The very existence of professional codes suggests that it is expected from the practitioners following them to have higher standards than those required by law. If somehow constraining, these codes are also massively approved and supported by professionals. A survey focusing on three types of professions – accountants, lawyers and engineers – has shown that in average, more than 80 % of the professionals thought it was necessary to have a code of conduct in a given profession (Higgs-Kleyn and Kapelianis 1999). What does this mean and how do these codes help professionals in their daily practice?

Professional Codes of Ethics embody what (Frankel 1989) calls the "society-profession nexus", and allow society to keep granting privileges to some professional bodies in return for their affirmed will to work in a way that is consistent with the society's core values. (Frankel 1989) identifies three main types of professional codes, while highlighting that these are often intertwined: aspirational – the code consists of "statements of ideals to which practitioners should strive"; educational – the code can provide helpful for dealing with specific situations – and regulatory – detailed rules that are enforced through a system of monitoring and sanctions. This profound hybrid nature of professional codes of ethics accounts for the multiplicity of their goals, as well as for many critiques they have received. In his seminal paper, (Frankel 1989) also identifies eight functions of professional codes of ethics. I will briefly outline each of these functions here as I believe a core issue of professional ethics arises from such a description.

1. *Enabling document*: gives the moral anchor to help the practitioner make more informed decisions
2. *Source of public evaluation*: functions as a basis for the public's expectations and evaluation
3. *Professional socialization*: gives a sense of group solidarity and common purpose
4. *Enhance profession's reputation and public trust*: helps persuade the public that professionals deserve its confidence and respect
5. *Preserve entrenched professional biases*: helps secure its members' professional monopoly
6. *Deterrent to unethical behavior*: creates a monitoring system supported by a range of sanctions
7. *Support system*: creates a legitimate support against improper demands
8. *Adjudication*: serves as a basis for adjudicating disputes between members or between members and outsiders

This multiplicity of functions suggests complex and intertwined relationships between three sets of players: the professional community, the individuals working inside the community, and society. Some functions form larger clusters that share a common objective, for instance, help the individual professional make a better decision. In its educational aspect, the code can then be seen as an enabling document;

professional socialization, allowing potentially fruitful discussions between members around shared objectives and challenges, arises from the aspirational perspective, while the regulatory aspect of the ethical code makes unethical behavior deterrent.

What we see however is that in a way, at least half of the functions listed here have as a primary goal the preservation of the professional community in itself and as it is, two directly (support system and preserve entrenched biases) and two indirectly by legitimizing its existence toward the general public (public evaluation and enhance profession's reputation and public trust). This observation is very similar to J.K. Galbraith's theories on the corporate world (Galbraith 1967) or to Ivan Illich's critique of institutions at large (Illich 1973): entities that become concerned only with their own survival, forgetting their initial purpose, whatever it may have been. From this perspective, there would be no wonder why practitioners massively support codes of ethics and codes of conduct: they support them before protecting society.

As their discipline grows, is this the path taken by GIS practitioners as well? We shall now see whether a specific code of ethics for GIS practitioners, the URISA code, intends to fulfill the same functions and whether or not it embodies the three types of ethical codes identified earlier – aspirational, educational, regulatory.

## 10.4 GIS Professional Ethics: Achievements and Limits

### *10.4.1 Do GIS Practitioners Need their Own Professional Ethics?*

According to (Chadwick 1998) and (Airaksinen 1994), in order to determine if a new group of professionals, such as GIS practitioners, should have their own professional ethics, this profession should meet three demands: (a) rely on a body of knowledge (b) and an ideal of service to the community (c) in which individual professionals have a high degree of autonomy in their practice. In the case of GIS, we can tick (a) without giving it too much thought: yes, GIS need a broad body of knowledge, both in geography and computer science. As for (b) and (c), it is less clear. Because of the equipment needed, GIS professionals usually work in teams, either for companies either for governmental organizations, and the level of autonomy of each individual inside these organizations may not be as important as the one of a doctor in private practice.

Let us now turn towards how GIS practitioners themselves define their profession and why they think they may need specific ethics. According to (Craig 2004), the reason why GIS practitioners need ethics is because "The GIS professional has many opportunities to do harm and to do good. We all try to make the right decisions, but sometimes it is not obvious what that decision should be". But this is not specific to GIS practitioners. In fact, the opportunities to do harm and good are rather less spontaneously obvious for GIS practitioners than they are for many other professions – even leaving apart the health care sector, one can probably think of professionals such as firemen, policemen, salesmen or teachers as having more daily opportunities to do harm and good than a person spending most of her time in front of a computer geo-coding spatial data. However, numerous professions have developed specific codes of ethics based on the same argument in the early 2000s (The main GIS Code of Ethics was published in 2003). (Gaumnitz and Lere 2002) for instance identify and analyze 15 Professional Business organizations that have their own code of ethics, among which accountants and realtors – we will see further down whether their codes of ethics shares similarities with the code of ethics developed for GIS practitioners.

The question whether an academic discipline or a field is being aware of the ethical issues it raises is no binary one. Rather, in an early attempt to categorize GIS ethics, (Crampton 1995) suggested that the notion of ethics a discipline holds can be considered as passing through four stages:

(a) Ignoring ethics (or rather being unaware of ethical issues)
(b) Considering ethics from an internal perspective only
(c) Considering ethics from both an internal and an external perspective
(d) Establishing a dialectical relationship, which modifies both internal and external perspectives. (Crampton 1995, p. 85)

The concept of "internal perspective", drawn from the work of geographer and cartographer Brian Harley, means here basically the concern for doing good science through sound scientific inquiry, while "external perspective" goes beyond the technical agenda and acknowledges the presence of an ideological context. Writing in 1995, J. Crampton considers that GIS Ethics have reached stage b indeed by focusing for instance on the need for "accurate" maps. Works by (O'Looney 1997) or (Esnard 1998) can be considered to belong to this same stage, by the focus they have on misuses and questionable uses of GIS and how to avoid them. The question needs to be asked whether, 15 years later, GIS Ethics have reached a new stage, or are still considering ethics within a purely internal perspective – thus not going much beyond the PAPA issues described above.

To address this question properly, it seems that a distinction needs to be drawn here between GIS academic ethics and GIS professional ethics. As GIS have evolved and developed new features (such as 3D visualization, real-time exploration), it seems that academic ethics have construed complex and reflective frameworks to accompany those changes. The work by (Sheppard and Cizek 2009) for instance, on the ethics of virtual globes software, is remarkable in its capacity to embrace an external, holistic perspective. However, the situation may not be the same when it comes to GIS Professional Ethics.

Geographers sometimes see themselves as a community which could need professional ethics, in a broad sense just as any academics could, about issues of data gathering and publication, and in a narrower sense anytime they deal with concepts such as space, place or nature, which is, all the time (Proctor 1998). But GIS present a specificity among geographers, which call for other needs:

> Ethical issues become more focused as one moves from a particular geographical concept to its technical implementation, and finally to its application. For instance, conceiving space as an isotropic surface appears innocent enough, until one builds a GIS upon this naïve assumption for the purpose of, say, specifying social service facility location. This example also suggests the interrelation of ethical issues across the continuum of geography's epistemological process, and points out the severe limitations in a "professional ethics" circumscribed solely to questions of research data and publication. (Proctor 1998, p. 14)

These specific needs call for a specific code of ethics for GIS practitioners.

### *10.4.2 The URISA Code of Ethics*

URISA (The Urban and Regional Information Systems Association), the broadest association for GIS Professionals, founded in 1963, acknowledged quite early the need for a code of ethics concerning GIS. The URISA Board of Directors therefore approved a Code of Ethics in April 2003, which is still in force. The introduction of the code clarifies its objectives: "This Code of Ethics is intended to provide guidelines for GIS (geographic information system) professionals. It should help professionals make appropriate and ethical choices. It should provide a basis for evaluating their work from an ethical point of view. By heeding this code, GIS professionals will help to preserve and enhance public trust in the discipline." (http://urisa.org/about/ethics). As duly noted by (Davis 2014) it is truly a professional code in the sense that it applies to all professionals and not only members of the Association, to the difference of organizational ethics – only applying to the members of the organization authoring a given code of ethics. This noteworthy difference highlights the universal take of the URISA code of ethics and the importance given to society as an entity. The code indeed identifies four types of obligations for GIS practitioners, towards different targets, and stating that obligations towards society should be preeminent compared to others.

1. **Obligations to society**: "The GIS professional recognizes the impact of his or her work on society as a whole, on subgroups of society including geographic or demographic minorities, on future generations, and inclusive of social, economic, environmental, or technical fields of endeavor." These obligations include the quality of the data ("do the best work possible"), their accessibility ("Contribute to the Community to the Extent Possible, Feasible, and Advisable"), and the communication around them ("Speak Out about Issues").
2. **Obligations to Employers and funders**: "The GIS professional recognizes that he or she has been hired to deliver needed products and services." Therefore, the URISA Code of Ethics recommends to "deliver quality work", "have a professional relationship", and "be honest in representations" – the latter having nothing to do with geographical or numerical representation, what one could have expected, but rather with the representation of the work: it goes from "deliver an hour's work for an hour's pay" to "describe services and products fully".

3. **Obligations to colleagues and the profession**: "The GIS professional recognizes the value of being part of a community of other professionals. Together, we support each other and add to the stature of the field." This passes through both respecting the work of others and contributing to the discipline to the extent possible (for instance by publishing the results so others can learn about them).
4. **Obligations to individuals in society**: "The GIS professional recognizes the impact of his or her work on individual people and will strive to avoid harm to them" by respecting individuals and their privacy, for instance by being "especially careful with new information discovered about an individual through GIS-based manipulations (such as geo-coding) or the combination of two or more databases."

Those four kinds of obligations seem rather exhaustive, and give the impression that the URISA Code of Ethics doesn't forget any community or any type of issues, therefore having reached at least stage c in the classification suggested by (Crampton 1995) described earlier. However, we will see that it is a moot point.

If we compare the URISA Code with codes from other professions established roughly at the same period, we see that some requirements are rather common, some original and some forgotten. In their study of 15 professional codes of ethics already mentioned earlier, (Gaumnitz and Lere 2002) identify nine major content categories in professional codes (see Table 10.1).

**Table 10.1** Main values in professional codes of ethics and how URISA relates to them

| Specific issue | % of professional codes mentioning them | URISA code of ethics |
|---|---|---|
| Confidentiality | 100 | Yes: "Hold information confidential unless authorized to release it." But in contradiction with other statements. |
| Honesty and integrity | 93 | Yes: "Practice integrity and not be unduly swayed by the demands of others" |
| Responsibilities to employers/clients | 93 | Yes: "Deliver quality work; Have a professional relationship; Be honest in representations". |
| Obligations to the profession | 87 | Yes: "Respect the work of others; contribute to the discipline to the extent possible" |
| Independence and/or objectivity | 80 | Yes: "Be objective, use due care, and make full use of education and skills." |
| Legal and/or technical compliance | 73 | No. |
| Discreditable or harmful acts | 67 | Yes: "The GIS professional recognizes the impact of his or her work on individual people and will strive to avoid harm to them." |
| Social values | 40 | Somehow |
| Ethical conflict resolution | 20 | Somehow |

The first remark here is that the only requirement that is present in all professional codes, the issue of confidentiality, does appear in the URISA code but seems to be in contradiction with other requirements. There is indeed a subsection in the URISA code calling for transparency under the title "speak out about issues". It can also be argued that the global aim of GIS practitioners is exactly the opposite of confidentiality: disclose and organize geo-coded data to make them available to the majority. The code of ethics actually also states, under "Contribute to the Community to the extent possible, feasible and advisable" "make data and findings widely available". But it also calls for respecting privacy at the individual level. In a word, the code seems of little help for GIS practitioners having to make a practical choice between disclosure and confidentiality of a specific set of data.

The following issues are present both in business professional organizations codes of ethics and in the URISA code: integrity, honesty, responsibility to employers, obligations to the profession, objectivity.

The legal and technical compliance, mentioned in the vast majority of professional business organizations' codes of ethics (73 %) is implicit in the URISA code, and is considered as something to go beyond: "Strive to do what is right, not just what is legal" (under "Do the best work possible").

"Social values" do not appear as such in the URISA code. What (Gaumnitz and Lere 2002) mean by social values is mostly the obligation to not discriminate, and obligations related to the public interest. The first one may be implied in the URISA code but is not phrased explicitly, and the second one falls under the general denomination of "society" without further specification. As for the resolution of ethical conflicts, several recommendations go in this direction but once again without being very specific.

Considering these limitations, it is therefore legitimate to wonder whether the URISA Code of Ethics is of much help for GIS practitioners in their daily practice.

### *10.4.3 The URISA Code of Ethics for GIS Practitioners: An Appropriate Charter?*

First of all, we cannot help but noticing, with (Craig 2004), that the URISA Code of Ethics only gives general guidelines, rather than specific rules of conduct. "For example, the text says things like "Be objective, use due care, and make full use of education and skills." It does not say things like which algorithm to use in a certain situation or whether a map legend is always required. The emphasis is on making GIS professionals aware of their actions and the impact of those actions." It is well understandable that a code cannot provide specific guidelines for all possible situations a practitioner may encounter throughout their career. It is in fact the very definition of a professional code, as an enabling document, that "a code may be compared to a compass, in that it provides the direction but does not presume to locate the ultimate decision" (Frankel 1989, p. 111). Here however the direction-providing aspect is questioned, as the given recommendations are so general that the very educational part does not seem fully fulfilled. But perhaps this is too much to ask from a code of ethics, if it cannot be based on a more general interest in professional ethics. (Huff 2014) insists on the importance of moral skills for GIS practitioners, as opposed to mere knowledge. He advocates for the inclusion of ethics in the academic curriculum of GIS practitioners, "not simply because we want them to mean well, we actually want them to do well in their profession". Maybe this would indeed be a way to make GIS codes of ethics more valuable in practice.

Another shortcoming of the code would be that no penalty is planned for those who would infringe the Code. Three causes are identified by Craig (2004) for the fact that the GIS Code of Ethics was presented without apparatus for penalizing those who are thought to violate it: First of all, "not having rules of conduct makes it difficult for a jury to judge whether a person is doing the right thing". Second, there is a concern over excessive resources spent on sanctions: the main job of a GIS Company is not to punish its employees for ethically questionable conducts. Third, there is another concern over potential anti-trust lawsuits coming from those whose earning capacity is reduced. Whatever cause is playing the most part here, the absence of monitoring and sanctions definitely takes away the regulatory aspect of professional codes of ethics identified earlier, thus leaving the URISA code to be mostly aspirational, perhaps educational to some extent, but in no sense regulatory.

The basic philosophy underlying the URISA code comes from the Kantian principle of always treating others with respect and never merely as means to an end. But an overwhelming ethical issue arises when there is a conflict between obligations to different entities, even though it is specified that "Obligations to society shall be paramount when there is conflict with other obligations": what about a conflict between the actions called for in the "obligations to society" and those about "individuals in society"? Who should get to decide what is more important then? Here it seems that the URISA Code fails to fulfill its adjudicatory function.

Finally, the only concrete advice formulated for GIS practitioners facing an ethical issue is to verbalize it and discuss it with colleagues: "The dialog with a colleague could help both of you. GIS professionals tend to think of themselves as competent people whose main concern is keeping current in the technology. We need to admit that we are part of the social world and reflect on the implications of our actions (or inaction). Then we will truly be professionals." concludes (Craig 2004). This focus on the decision-making process rather than normatively spelling out desirable outcomes somehow comes close to Habermas' discourse ethics. But Habermas precisely did not write any code supposedly helping out individuals or community making tough decisions, his work is rather concerned with the very possibility of communication and mutual understanding. Perhaps is this not enough of an ambitious objective for a professional code?

## 10.5 Conclusion: Rethinking the Expert-Lay Interplay for GIS

*In fine*, what (Craig 2004) recommends for GIS practitioners to do in order to be truly professionals, is to dialogue with their colleagues, to engage in debate with other members of society, which is, a democratic process. Now, society as a whole might benefit from these interoffice talks, but shouldn't it be a part of it? We have established that many of the ethical challenges raised by GIS depend on the political and social context in which GIS are implemented and used. Shouldn't it be then society, and probably GIS companies, as part of society, who should discuss what *we*, lay-persons, GIS users, common citizens, owners of houses which might end up in a GIS someday, lost tourists using Googlemaps to find our way, truly want? Do *we* value our privacy more than the possibility of using geo-referenced maps? Do *we* prefer to pay in order to protect our personal data, or maybe have more accurate information? Or do we prefer to spend time in participative GIS so that there is a shared ownership of the geospatial data we would contribute to? The significance of these individual choices is indirectly highlighted by Turner (2006) and Harvey (2014) when they remark that the development of mobile phones "will hold far more extensive impacts for spatial technology users than today's $15B/year GIS industry" (Harvey 2014, p. 501). As expressed in another paper of this volume about the ethical issues raised by the modifications of the human body, the whole question is then about "What is the master value of our society?" (Holm 2016, in this volume), or rather in plural: what are the values that we take responsibility for, or are willing to make sacrifice for?

It appears then that GIS are another illustration of the Kranzberg law, stating that technology is deeply embedded in social and cultural contexts, so will be its ethical developments, for the good or for the bad. To a large extent, ethical challenges of using GIS in democratic societies arise not only and probably not mainly from the specific technical characteristics of GIS, but from democracy in itself: wherever a democratic system is trustworthy and reliable, both by promoting reliable democratic institutions and by including citizens in decision-making processes, GIS could develop into a legitimated and trustworthy tool for representing space-society relations. Within non-democratic settings, we couldn't help it being a threat to individuals and society in its whole.

Naturally, a further challenge presented by GIS is that it's a global phenomenon, in every considered sense: GIS do map the earth in its whole, including oceans and the atmosphere, GIS companies are based in many countries, employing professionals from all origins. It is then very unlikely that we will have in a close future a political tool suited to address the ethical issues raised by a worldwide technology mapping and framing the territory all human beings, with their different and mostly incompatible values systems, live in. Failing which, I would like to suggest at least an individual rule of self-consistency: if *you*, as an individual citizen, are disturbed by the existence and spread of GIS such as Google Earth or Google Street View, before going to protest in the streets of Berlin, or before signing petitions to powerless political institutions, a necessary, although maybe not always sufficient first step is to stop using them, downloading them, enjoying them. GIS would not exist without their users. But if *you* don't mind having your street or even your house visible by all Internet users, well then keep enjoying GIS, with a clear conscience.

## References


Airaksinen, Timo. 1994. Service and science in professional life. In *Ethics and the professions*, ed. Ruth Chadwick, 1–13. Aldershot: Avebury.

Barndt, Michael. 2002. A model for evaluating public participation GIS. In *Community participation and geographic information systems*, ed. William J. Craig, Trevor M. Harris, and Daniel Weiner, 346–356. London: Taylor & Francis.

Böhm, Gisela, and Hans-Rüdiger Pfister. 2000. Action tendencies and characteristics of environmental risks. *Acta Psychologica* 104: 317–337.

Brown, Rosemary. 1965. *Professional codes of ethics: A study of the judicial viability of the codes of ethics of medicine, social work and librarianship*. School of Social Work: University of British Columbia.

Chadwick, Ruth. 1998. Professional ethics. In *Routledge encyclopedia of philosophy*, ed. Edward Craig. London: Routledge. http://www.rep.routledge.com/article/L077. Accessed 14 Feb 2011.

Chrisman, Nicholas. 2006. Full circle: More than just social implications of GIS. *Cartographica* 40: 23–35.

Craig, William J. 2004. GIS ethics: Understanding implications of action. *GIS Development*. http://www.gisdevelopment.net/magazine/years/2004/July/14.asp. Accessed 11 Feb 2011.

Crampton, Jeremy. 1995. The ethics of GIS. *Cartography and Geographic Information Systems* 22: 84–89.

Curry, Michael R. 1995. Geographic information systems and the inevitability of ethical inconsistency. In *Ground truth: The social implication of geographic information systems*, ed. John Pickles, 68–87. New York: Guilford.

Davis, M. 2014. What to consider when preparing a model core curriculum for GIS ethics: Objectives, methods, and a sketch of content. *Journal of Geography in Higher Education* 38(4): 471–480. doi:10.1080/03098265.2014.956298.

Esnard, Ann-Margaret. 1998. Cities, GIS and ethics. *Journal of Urban Technology* 5: 33–45.

Frankel, M. 1989. Professional codes: Why, how, and with what impact? *Journal of Business Ethics* 8: 109–115.

Galbraith, John Kenneth. 1967 (2007). *The new industrial state*. Princeton: Princeton University Press.

Gaumnitz, B., and J. Lere. 2002. Contents of codes of ethics of professional business organizations in the United States. *Journal of Business Ethics* 35(1): 35–49. doi:10.1023/A:1012718103007.

Gewin, Virginia. 2004. Mapping opportunities. *Nature* 427: 376–377.

Gorman, Sean. 2011. The issues of scope and scale that face GIS. *GeoIG*. http://blog.geoiq.com/2011/06/07/the-challenges-of-scope-and-scale-that-face-gis. Accessed 11 June 2011.

Haque, Akhlaque. 2003. Information technology, GIS and democratic values: Ethical implications for IT professionals in public service. *Ethics and Information Technology* 5: 39–48.

Harvey, F. 2014. Values, choices, responsibilities: Thinking beyond the scholarly place of ethics for the GIScience and technology profession and GIScience. *Journal of Geography in Higher Education* 38(4): 500–510. doi:10.1080/03098265.2014.956299.

Higgs-Kleyn, N., and D. Kapelianis. 1999. The role of professional codes in regarding ethical conduct. *Journal of Business Ethics* 19(4): 363–374. doi:10.1023/A:1005899517191.

Holm, Søren. 2016. The modification of the human body: Controversies. In *Technoscience and citizenship: Ethics and governance in the digital society*, ed. Ana Delgado. Cham: Springer.
Huff, C. 2014. From meaning well to doing well: Ethical expertise in the GIS domain. *Journal of Geography in Higher Education* 38(4): 455–470. doi:10.1080/03098265.2014.936314.
Illich, Ivan. 1973. *Tools for conviviality*. New York: Harper & Rows.
Jepsen, Sune, Delgado A. Kim, and M. Bertilsson. (forthcoming). Sparkling publics into being: Reflections on public engagement with technology in contemporary democracies.
Longhorn, Roger A. 2004. Identifying ethical and unethical activities in GIS. *GIS Development*. http://www.gisdevelopment.net/magazine/years/2004/July/18.asp. Accessed 11 Feb 2011.
Mason, Richard O. 1986. Four ethical issues of the information age. *MIS Quarterly* 10: 5–12.
O'Looney, John A. 1997. *Beyond maps: GIS and decision making in local government*. Washington, DC: ICMA.
Proctor, James D. 1998. Ethics in geography, giving moral form to the geographic imagination. *Area* 30: 8–18.
Russell, Bertrand. 1954 (2010). *Human society in ethics and politics*. London: Routledge Classics.
Sheppard, Stephen R.J., and Petr Cizek. 2009. The ethics of Google Earth: Crossing thresholds from spatial data to landscape visualization. *Journal of Environmental Management* 90: 2102–2117.
Turner, A. 2006. *Introduction to neogeography*. Sebastopol: O'Reilly Media.
Winner, Langdon. 1977. *Autonomous technology: Technics-out-of-control as a theme in political thought*. Cambridge: MIT Press.
Winner, Landgdon. 1986. *The whale and the reactor: A search for limits in an age of high technology*. Chicago: University of Chicago Press.

# Chapter 11
# Geoengineering: Reflections on Current Debates

**Paula Curvelo and Ângela Guimarães Pereira**

**Abstract** In this paper we propose to investigate the current debates on geoengineering, here considered as an illustrative metaphor of particular technoscientific promises and 'techno-fix' narratives that are emerging in our society. After a brief introduction, where we provide the necessary background to understand the complex issues surrounding geoengineering, we discuss the relevance of this investigation. We then proceed to explore the controversies behind geoengineering, which start with its own definition. The analysis of the current debates around geoengineering experimentation, regulation and deployment reveal some of the dominant narratives of technoscientific progress and highlight important tensions and frictions in the relationship between science, policy and society. A reflection on these issues suggests the relevance of developing alternative approaches to furthering the 'democratisation and de-alienation' of geoengineering debates, thus responding to a perceived need for more careful consideration of the normative assumptions that lie behind the idea of deliberately manipulating Earth's climate to offset anthropogenic climate change.

**Keywords** Climate Engineering • Geoengineering Debates • Climate Change • Narrative Inquiry • Master Narratives

## 11.1 Introduction

The idea of weather modification and climate manipulation is not new, nor is it associated with a particular discipline, branch of knowledge or area of expertise. For this reason, the (hi)story of geoengineering – i.e. 'the deliberate large-scale intervention in the Earth's climate system in order to moderate global warming' (The Royal Society 2009) – may have different beginnings, may explore (or omit) a variety of

The views expressed in this paper are those of the authors and do not necessarily represent the official views of the European Commission.

P. Curvelo (✉) • Â. Guimarães Pereira
European Commission – Joint Research Centre, Ispra, Italy
e-mail: paula.curvelo@gmail.com



A. Delgado (ed.), *Technoscience and Citizenship: Ethics and Governance in the Digital Society*, The International Library of Ethics, Law and Technology 17,
DOI 10.1007/978-3-319-32414-2_11

episodes and may be told from a particular point of view. These 'qualities' are visible in the fairly recent literature on the topic, where we find a variety of references introducing and contextualising the emergence of geoengineering technologies: the image of Ulysses assisted by, or being a victim of, deliberate weather modification schemes brought about by various gods and goddesses; the tempest conjured up by Prospero in William Shakespeare's play of the same name (Schneider 1996); the diabolical plan to tilt the Earth's axis and melt the polar ice in Jules Verne's novel *The Purchase of the North Pole* (Fleming 2006); the various attempts at artificial rainmaking (Fountain 2003; Fleming 2007, 2010), in what Robert DeCourcy Ward called the stage of 'production' (Ward 1930); the weather control fantasies of military planners and the way these visions shaped some of the weather modification programmes in the Cold War era (Bonnheim 2010; Fleming 2006, 2010; Keith 2000); the common links with the concept of 'terraforming' (Fleming 2010; Yanarella and Rice 2011) and the way its literature (both scientific and fictional) is trying to fill the gaps that still exist in geoengineering literature (Keith 2000).

These are but a few examples of the ubiquitous and eclectic ideas that, in some way, share a common theme with current geoengineering proposals: that which relates human expectations, fears and fantasies with the recurring appeal of the control of nature. Hence, even though some may appear unconnected with the history of geoengineering, such ideas are nonetheless useful in reminding us that recent proposals to geoengineer the climate are just one contemporary manifestation of man's long-standing desire to control nature – an early-twenty-first-century embodiment of the 'Baconian project' of human mastery over nature.

By looking at 'the long history of deceptive and delusional attempts to control nature', Fleming identifies three cycles of 'promise and hype' that capture the pathological features of weather and climate control schemes (Fleming 2006, 2007).

The first cycle, the 'Pluviculturalists', began in the 1840s with the work of the meteorologist James Pollard Espy, who propounded a theory of artificial rainmaking by lighting huge fires.

The second cycle, 'Cloud seeding in the Cold War and Vietnam War eras', began in 1946 with the pioneer experiments in cloud seeding by Irving Langmuir and his associates at the General Electric Research Laboratory, which rapidly evolved from lab science experiments to commercial rainmaking applications, and ultimately 'the attempted weaponization of the clouds' (Fleming 2006).

Three decades later, the term geoengineering was coined by the physicist Cesare Marchetti to describe a proposal for tackling the problem of $CO_2$ control in the atmosphere with a $CO_2$ management system, where '$CO_2$ is collected at proper fuel transformation points and finally injected into the deep seas taking advantage of natural thermohaline circulations' (Marchetti 1977). Almost at the same time, on the other side of the globe, the Russian climatologist Mikhail Budyko was probing the potential of different techniques to modify the aerosol layer of the stratosphere to prevent the warming of the climate (Bonnheim 2010; Budyko 1977; Schneider 1996).

However, it was only at the beginning of this century that geoengineering entered the mainstream debate on climate change. According to Fleming, the beginning of the third cycle, 'Weather modification in the 21st century' – in which 'discussion of weather and climate modification has returned to the science-policy agenda, framed as seemingly inevitable responses to killer storms and global warming' – coincides with the publication of the U.S. National Research Council report titled "Critical Issues in Weather Modification Research" (National Research Council 2003), and the report commissioned by the U.S. Pentagon, "An Abrupt Climate Change Scenario and its Implications for United States National Security" (Schwartz and Randall 2003). But perhaps the most important impetus came in 2006, with the publication of an editorial essay by Nobel laureate Paul Crutzen in the journal *Climatic Change* (Crutzen 2006) that brought discussions of geoengineering more squarely into the focus of scientific debates (Pielke, Roger Jr. 2010).

Yet, to fully understand these events, we have to consider them in the context of increasing doubt and disbelief regarding the commitment of the international community to adequately respond to the problem of global warming. In fact, throughout the twenty-first century, the geoengineering discourse has been closely coupled with the climate change agenda, being affected by its major convulsions in the scientific and political arenas.

Lastly, the continued misunderstanding and disbelief in the science of climate change, the recognition that global warming is the net result of various institutional failures, and the recent tendency to favour transformational (rather than incremental) responses to this problem, appear to have combined with the major uncertainties of climate change to provide the conditions for geoengineering to emerge as a paradigmatic case 'where facts are uncertain, values in dispute, stakes high and decisions urgent' (Funtowicz and Ravetz 1993, 1994a).

## 11.2 Exploring Geoengineering Debates

Though the idea of weather and climate control is not new, the purpose and extent of climate modification proposals since the beginning of this century seem to have overtaken the original concepts and the scientific questions from which they arose, and have been appropriated by the competing interests that surround climate change science.

The environmental problems and scientific uncertainties that many of the climate engineering schemes evoke are being brought to the centre of the climate change debate, feeding environmental controversies and bringing to light value disputes at the same time as the discourse becomes more and more politicised (Sarewitz 2004).

Against this background, the analysis of geoengineering debates may contribute not only to uncovering the variety of knowledge, values and interests that compete in the climate change science, but also to mapping the dynamics of these debates in the context of the major narratives that are emerging in our society—thus seen as a valuable approach to understanding the mutual co-production of science and society, in which 'scientific knowledge both embeds and is embedded in social identities, institutions, representations and discourses' (Jasanoff 2004).

In this context, approaching geoengineering in a holistic manner is another way of looking at the problem of climate change and the 'scalar dislocations' it introduces in modern systems of experience and understanding (Jasanoff 2010). In fact, the ethical, political, environmental and social considerations that surround the debates on geoengineering seem to offer a privileged perspective for rethinking the human place in nature.

### *11.2.1 Debates Around the Definition of Geoengineering*

We start the analysis of current debates on geoengineering by focusing on the major disputes around the definition of geoengineering – a term on which the scientific community seems far from reaching a consensus, as has been pointed out by several authors and was made clear in the 2011 IPCC expert meeting on geoengineering.

> A substantial amount of time in the Expert Meeting was spent in discussing terminology in and around geoengineering. This underlines the ambiguities associated with the term geoengineering and the range of opinions on the subject (Boucher et al. 2011, p. 2).

In fact, many of the controversies surrounding geoengineering start with the lack of consensus regarding the broadness and significance of the term. A look at the recent literature on the topic (scientific articles, books, policy reports and media articles) reveals two major sources of disagreement. The first of these is the different meanings attributed to the term. One example of this may be seen in the confrontation between those authors that suggest that we began geoengineering the Earth's climate when we started causing significant disturbances to the planetary environment (resulting in a definition of geoengineering closely related to that of the 'Anthropocene'[1] (Crutzen and Stoermer 2000)), and those espousing definitions that highlight the particular characteristics of the actions carried out with different climate engineering techniques.

> In its broadest sense, geoengineering involves deliberately modifying the Earth system and its processes to suit societal needs and improve the planet's habitability. During recent years, discussions of this controversial concept have been confined largely to global-scale engineering approaches intended to counteract the effects of anthropogenic climate change. Proponents of geoengineering point out that humans have been modifying the Earth system and its processes unintentionally for some time; therefore, why not do it in a deliberate manner with specific goals in mind? (Greene et al. 2010).

As the above quotation suggests, a broad definition of geoengineering tends to underestimate the arguments against the most controversial schemes to modify the energy balance of the atmosphere. By contrast, a narrow definition of the term highlights the intentionality of geoengineering actions, thus calling for a critical examination of the ethical, social, and political issues raised by these proposals.

But even if general agreement could be achieved on the particularities of geoengineering actions, the different types of proposals that the term encompasses seem

[1] Paul Crutzen and Eugene Stoermer coined the term *Anthropocene* to describe a new geological epoch, '*in which humankind has emerged as a globally significant—and potentially intelligent—force capable of reshaping the face of the planet*' (Clark et al. 2004) .

to be a second source of fuzziness more difficult to address. To illustrate this, we briefly present the two different families of methods into which geoengineering schemes are usually classified (The Royal Society 2009):

(i) Carbon Dioxide Removal (CDR) methods, which aim to reduce the concentration of $CO_2$ in the atmosphere and transfer it to long-lived reservoirs, and
(ii) Solar Radiation Management (SRM) methods, which aim to reduce the amount of solar energy absorbed by the Earth.

The first family of methods includes large-scale engineering approaches, which use either chemical or physical processes to directly remove $CO_2$ from the atmosphere or the oceans, (e.g. engineered air capture and enhanced weathering techniques), and biologically-based methods seeking to simulate or enhance natural carbon storage processes (e.g. afforestation and reforestation, biomass and biochar, ocean fertilisation methods, among others).

The second family of methods includes some of the most controversial geoengineering proposals. Four major groups of techniques have been proposed to reduce the incidence and absorption of incoming solar radiation: (i) *Space-based approaches* – reducing the amount of solar energy reaching the Earth by positioning sun-shields in space with the aim of reflecting or deflecting solar radiation; (ii) *Changes in stratospheric aerosols* – injecting sulphates or other types of particles into the upper atmosphere, with the aim of increasing the scattering of sunlight back to space; (iii) *Increases in cloud reflectivity* – increasing the concentration of cloud-condensation nuclei in the lower atmosphere, particularly over ocean areas, thereby whitening clouds with the aim of increasing the reflection of solar radiation; and (iv) *Increases in surface albedo* – modifying land or ocean surfaces with the aim of reflecting more solar radiation out to space (The Royal Society 2009; Williamson et al. 2012, p. 26).

As this brief overview suggests, the different technological characteristics of these proposals, the different costs estimated for each method, the potential efficacy of their use, the levels of uncertainty associated with their deployment, and the distinctive risks they raise result in a multitude of solutions that seem difficult to bring together under the broad umbrella of the term geoengineering.

One of the first attempts to clarify the ambiguity of the term was made in 1996 by Thomas Schelling, who identified the features that geoengineering seems to imply: global, intentional and unnatural interventions (Schelling 1996). Four years later, David Keith took this proposal further by pointing to the three core attributes that serve as 'markers of geoengineering' actions: the scale (global or continental), the intent (the deliberate nature of the action rather than a side effect of it) and the degree to which the action is a countervailing measure (Keith 2000). The reasons for replacing 'unnatural' features by the 'degree to which the action is a countervailing measure' were not properly explained – even though this had implications for the type of proposals that the term encompasses.[2]

[2] In order to exemplify these implications it seems appropriate to refer briefly to the use of weather modification techniques (such as cloud seeding and hurricane suppression) that are taking place in

Nevertheless, these three markers seem to translate the meaning of the term geoengineering as commonly used by the scientific community nowadays, furthering the conceptual distinction between geoengineering proposals and other responses to climate change.

However, in considering the ethical issues raised by these technologies, it becomes clear that these markers tend to hinder the various values, rationales and normative assumptions underlying the range of CDR and SRM techniques considered under the broad umbrella of the term geoengineering. As mentioned by Gardiner, the ethical discussion of geoengineering is made more difficult by the complexity of the terrain:

> First, a number of interventions are already being proposed for combating climate change, and it is not clear that all of them should be classified together. For example, some suggest deflecting a small percentage of incoming radiation from the Sun by placing huge mirrors at the Legrange point between it and the Earth, some advocate fertilizing the oceans with plant life to soak up more carbon dioxide, some suggest a massive program of reforestation, and some propose capturing vast quantities of emissions from power plants and burying them in sedimentary rock deep underground. But do these interventions raise the same issues? Should we count all of them as "geoengineering"? (Gardiner 2010, p. 285).

To overcome the obstacles raised by the broadness of the term, Bunzl appeals to the methodological distinction between small 'g' proposals and big 'G' proposals.[3] According to the author, this distinction is fundamental to deconstructing some of the common arguments for advancing further and faster in geoengineering research. In fact, because big 'G' proposals fall into a specific class of scientific endeavours (where the object of interest is not 'modular' or 'encapsulated'), they generate a set

many countries around the world. As recently stated by the World Meteorological Organisation: '*since the 1980's there has been a decline in support for weather modification research, and a tendency to move directly into operational projects*' (WMO 2010). Given the similarities between weather modification (WM) techniques and some geoengineering methods, the concerns raised by the increasing number of WM operational programmes (fog dispersion, rain and snow enhancement and hail suppression) have gained momentum in the context of the contemporary debates on geoengineering—leading, almost inevitably, to a discussion on the criteria that differentiate these two domains. However, although widely mentioned, the scale marker seems to be insufficient to exclude WM techniques from the vast range of methods that the term geoengineering encompasses. This becomes clear from the way the 'countervailing measure criterion' has been evoked, namely by drawing attention to the differences between 'weather' and 'climate' modification techniques, and to the far-reaching consequences of the latter: '*Weather modifications such as cloud seeding which affect the weather for no longer than a season, in our view, do not fall within the definition of geoengineering (…) We conclude that weather techniques such as cloud seeding should not be included within the definition of geoengineering used for the purposes of activities designed to effect a change in the global climate with the aim of minimising or reversing anthropogenic climate change*' (UK House of Commons. Science and Technology Committee 2010, p. 15). Nonetheless, it is telling that according to this same report: '*Cloud seeding could affect climate when carried out over a long period*' (Idem).

[3] '*Of course there is geoengineering and then there is GEOENGINEERING. Nobody gets very wound up about the idea of planting trees or painting roofs white as instances of geoengineering—which is not to say that they will necessarily do much good. The kind of geoengineering that elicits howls of disapproval is grander than this—it is things like space mirrors, sulphur injection into the upper atmosphere, and iron fertilisation of the oceans—it is the idea of intervention on a grand scale*' (Bunzl 2009).

of methodological challenges, allowing the moral argument as to 'whether research should be done' to give way to the methodological argument as to 'whether it could be done'—thus shifting the burden of proof to the proponents of geoengineering.

> But what if the object of your interest is not modular or encapsulated? What do you do then? For that, after all, is the feature that big "G" geoengineering proposals have in common. They call for interventions on systems that lack just this characteristic. You cannot encapsulate part of the atmosphere and it is too complex to be able to build a realistic non-virtual model at scale. As such, it is reasonable to ask whether we could ever have a sound basis for moving to full deployment of any such proposed intervention. And if not, then why bother to even research such proposals in the first place? (Bunzl 2009, p. 2).

It seems most reasonable to question the feasibility of geoengineering research in light of its object of interest. Indeed, the pressure of practice under which science operates today (Carrier 2011) is giving rise to the emergence of new objects of research – ambivalent beings, hybrid products and theoretically constructed objects through which we gain a new understanding and control of nature – that call for a more careful consideration of the complex narratives and practices of science and technology (Latour 1987; Funtowicz and Ravetz 1993, 1994a, b; Haraway 1997; Law 2002; Michael 2006).

Following this appeal, some authors have suggested that it is precisely at the level of these objects of research that we can find the meaningful distinction between science and technoscience, an ontological difference that 'becomes more explicit when research results are presented in particular settings and when the objects of research are exhibited for the specific interest they hold' (Bensaude-Vincent et al. 2011, p. 365). Accordingly, and by way of illustration, it could be said that when the result of a global climate model experiment is presented as scientific evidence for understanding the role of aerosols in climate forcing, this would conform to traditional conceptions of science. However, when sulphate aerosols are presented for their capacity to counteract the climate forcing of growing $CO_2$ emissions, this should be seen as a 'hallmark of technoscience'.

As we will see next, many of the controversies surrounding geoengineering go beyond the ambiguities of the term. However, and as pointed by Bunzl, some important questions regarding the research, governance and deployment of these technologies can only be properly answered if we consider the significant differences between the multitude of proposals that the term encompasses.

## 11.3 Unveiling the Multiple Narratives Behind Geoengineering Discourses

An overview of the subject of geoengineering may be extremely illustrative, but also bewildering, with regard to the controversies that the term encompasses. At first glance, one may be frightened by the revisited version of Edvard Munch's 'The Scream' on the cover of the report of the ETC Group, entitled 'Geopiracy – The Case Against Geoengineering' (ETC Group 2010). Perhaps one may also find this reference in a suggested comparison between the way the Krakatoa eruption inspired Munch to create this work and the way the eruption of Pinatubo inspired geoengineers to cool the Earth (Hamilton 2010). One may come across the variety of histories that feed 'chemtrails' theories,[4] or one of the scientific studies that compare different geoengineering options (Keith and Dowlatabadi 1992; Keith 2000; National Research Council 1992; Vaughan and Lenton 2011). By chance, one may stumble over a few of the various attempts to rank these options: some evocative but difficult to assess (Adam 2009), others suggesting a scientific asset (Boyd 2008; Lenton and Vaughan 2009; The Royal Society 2009), and yet others being sarcastic about this last possibility (Singer-Vine 2010). Lastly, one will most likely end up with Meinrat Andreae's decadent image of our society's addiction to fossil fuels, 'It's like a junkie figuring out new ways of stealing from his children' (Morton 2007), or with one of the many meaningful terms that populate the geoengineering world: 'back-up plan' (Inman 2010); 'catastrophic climate change'(Gardiner 2011; Hegerl and Solomon 2009); 'climate anxiety' (Bonnheim 2010); 'covert geoengineering' (Lawrence 2006); 'emergency brake' (Brovkin et al. 2009); 'fallback strategy' (Keith 2002; Keith and Dowlatabadi 1992); 'geohack' (Singer-Vine 2010); 'global thermostat' (Goodell 2010); 'planet-hacking techniques' (Kintisch 2010); 'planetary medicine' (Lovelock 2008, 2009); 'predatory geoengineering' (Gardiner 2011); 'retooling the planet' (Bronson et al. 2009); 'stopgap' (Barrett 2008; Bunzl 2009); 'technological fix' (Montenegro and Greenwood 2009).

Despite the confusion a first glance may suggest, a more detailed analysis of the literature in the field may be extremely valuable in understanding the particular kinds of knowledge, values and interests that are competing in the climate change debate and in uncovering some of the dominant narratives that operate at different levels of society.

One reasonable and logical way of digging through the debates on geoengineering is 'to collect, structure, and relate the very different arguments that have been advanced for and against climate engineering' (Rickels et al. 2011): the 'moral hazard argument' (The Royal Society 2009); the 'slippery slope' argument; the 'technical fix' argument; the 'unpredictability' argument (Keith 2000); the 'lesser evil' argument; the 'arm the future' argument, the 'cost-effectiveness' argument; the 'research first' argument; the 'stalking horse' argument (Gardiner 2010, 2011); the 'common sense' argument (Jamieson 1996); the 'desperation argument' (Gardiner 2012), etc. These different arguments tend to be linked around the main theses that have been identified in the debate on the pros and cons of geoengineering research and deployment (Betz and Cacean 2012; Rickels et al. 2011) and illustrate the diversity of attempts in this area. Many of these arguments emerge in the debates in and around the 'grey zones', or interfaces, between science, policy and society (Siune et al. 2009), and can be grouped, for systematisation purposes, into three interconnected domains: (i) geoengineering research and experimentation; (ii) geoengineer-

[4] The term 'chemtrail' is derived from 'chemical trail' and specifically refers to chemical or biological agent trails left by aircraft for a purpose undisclosed to the general public, allegedly causing respiratory illnesses and other health problems.

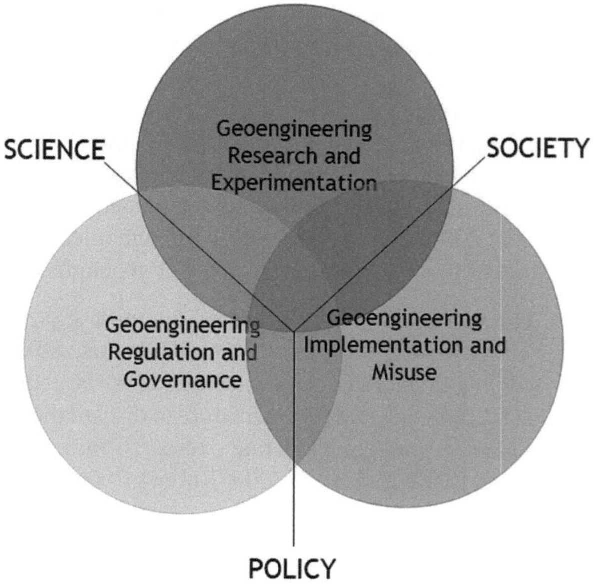


**Fig. 11.1** Domains within which current debates on geoengineering can be grouped

ing regulation and governance, and (iii) geoengineering implementation and misuse (Fig. 11.1).

### *11.3.1 Geoengineering Research and Experimentation*

In this first domain, we have identified some of the most active disputes over geoengineering, which is not surprising given that research is the stage where much of the geoengineering proposals currently are at and experimentation is the expected next step. The construction of the arguments varies, but in general they fall into three major groups:

(i) The first group holds that geoengineering, along with mitigation and adaptation, is a valid and unavoidable response to climate change, so we must invest in geoengineering research in order to be prepared for a likely climate emergency:

> The rate of increase of climate change, along with the continuing increase in emissions of greenhouse gases, has created a very serious predicament for the world. Drastically reducing the world's use of fossil fuels will take time and may raise near-term costs for energy, even after the effort gets seriously started and production costs for new energy technologies drop. As a result, global warming is likely to press up against or even exceed the level that the Commission of European Communities, for example, has concluded is likely to lead to dangerous and unacceptable consequences. For

this reason, it seems prudent for the nations of the world to initiate an effort in geoengineering (…) (MacCracken 2009, p. 33).

This argument finds its support in two inter-related narratives of technoscientific progress that tend to shape and frame key dimensions of science and governance: the risk of dismissing a 'promising technology' and the 'speed imperative' that impels us to act immediately 'before it is too late' (Felt et al. 2007). Furthermore, this argument often appears coupled with another, that of the 'incredible economics of geoengineering' (Barrett 2008).

(ii) The second group, and perhaps the most significant, regards geoengineering with reserve but considers it would be a mistake to ban geoengineering research without first reducing the uncertainties surrounding the associated benefits and risks, claiming that it is premature to discard these options without carrying out adequate, though 'moderate', research into the topic (Blackstock and Long 2010; Blackstock et al. 2009; Robock 2008, 2011).

The reasons why geoengineering may be a bad idea are manifold, though a moderate investment in theoretical geoengineering research might help scientists to determine whether or not it is a bad idea (Robock 2008).
This strong argument is sustained by the new credo of 'evidence-based' decision making, where facts must precede any exercise of values.

(iii) The third group comprises arguments against geoengineering research – particularly research into SRM methods. The arguments are of two kinds:

- those opposing geoengineering solutions on principle, on the basis of the common sense belief that 'two wrongs do not make a right', and
- those challenging the soundness of the arguments in favour of geoengineering research, in this way trying to deconstruct some of the narratives on which they are based.

Since the first kind of argument concerns the validity of geoengineering solutions, it does not directly address the specific case of geoengineering research (which is assumed to be as doubtful as the concept itself). Here, we can find many of the arguments that try to demolish the 'techno-fix' ideas behind geoengineering solutions: SRM methods do not address the root cause of anthropogenic climate change; geoengineering is 'unnatural' and SRM technologies are objectionable (Jamieson 1996; NERC 2010a, b); we have to find a place consistent with the limits of nature (Bunzl 2009) and technology cannot replace the process of 'social engineering' (Weinberg 1991) that this goal implies.

The second kind of argument against geoengineering research tends to emerge in the ethical discussion of the subject (Keith 2000; Gardiner 2011; Bunzl 2009; Victor et al. 2009; Jamieson 1996; Hamilton 2013). Morrow, Kopp and Oppenheimer present a clear synthesis of these arguments:

> There are four ethical reasons to worry about performing climatic SWCE[5] research at all, over and above its effects on humans, animals, and ecosystems. First, pursuing SWCE solutions to climate change may create a moral hazard, exacerbating the challenge of mitigating emissions. Second, SWCE research may lead to development of technologies that could be used for nefarious purposes. Third, beginning SWCE research in earnest may create interest groups within scientific or business communities that would have strong incentives to push for SWCE (or at least SWCE research) even if it turns out to be unwise. Finally, money spent on SWCE research is unavailable for other kinds of research, such as on the mitigation of or adaptation to climate change (Morrow et al. 2009).

Discussions about governance mechanisms and basic principles to guide future geoengineering research tend to highlight the profound reorientation of technoscientific practices in contemporary societies – of what has been seen as a major shift from the 'laboratory ideal' to the 'field ideal' of experimentation (Schwarz and Krohn 2011). Indeed, the recognition that 'several geoengineering technologies are demonstrably non-encapsulated' (Bracmort and Lattanzio 2013, p. 5) tends to further polarise the debate. On the one hand, those who call for 'a moratorium on all geoengineering activities outside the laboratory' (ETC Group 2010, p. 40). On the other hand, those who consider that the 'least risky option would involve starting with small-scale field experiments and gradually ramping up the scale' (Eccleston and March 2011, p. 358).

> Another key question is how to address further research. Proponents of further research argue that it is needed in order to obtain reliable information about the feasibility and risks. However, this would at some stage require real-world field experiments that would have to be gradually scaled up in order to know the impacts of a particular technique and whether it is effective. Apart from the difficulty of drawing the line between research and deployment, most existing rules of international law do not make this distinction (Bodle 2013, p. 468).

And once again, Morrow, Kopp and Oppenheimer provide an interesting point of view on this subject, introducing the 'narrative of ethics' to the debate, and suggesting that climatic SWCE research[6] is very similar to nuclear weapons testing. They thus propose careful ethical consideration guided by three principles derived from the ethics literature on research with human and animal subjects:

> The Principle of Respect requires that the scientific community secure the global public's consent, which would need to be voiced through their representatives and given for any studies within specified parameters, rather than on a case-by-case basis. The Principle of Beneficence and Justice requires that researchers strive for a favorable risk-benefit ratio and a fair distribution of risks and anticipated benefits, all while protecting the basic rights of the individuals affected. Finally, the Principle of Minimization requires that no study last longer, cover a greater geographical extent, or exert a greater influence on the climate than is necessary to test the specific hypotheses in question (Morrow et al. 2009, p. 1).

---

[5] In this paper the authors refer to SRM as "short-wave climate engineering" (SWCE).

[6] According to the authors, other kinds of climate engineering research (such as modelling studies and engineering studies) do not raise the same concerns as climatic studies – which aim '*to determine the climatic response to climate engineering and therefore could have widespread impacts on both human populations and the biosphere*'.

A similar position was articulated in the "Oxford Principles", a set of five overarching principles for governance of geoengineering research: (i) geoengineering to be regulated as a public good; (ii) public participation in geoengineering decision-making; (iii) disclosure of geoengineering research and open publication of results; (iv) independent assessment of impacts; and (v) governance before deployment (Rayner et al. 2009; Rayner et al. 2013). This leads us to the second domain of the geoengineering debate: that of geoengineering regulation and governance.

### *11.3.2 Geoengineering Regulation and Governance*

In this second domain, the debate revolves around two main concerns: the need (i) to regulate specific geoengineering activities (large-scale research projects, small-scale field tests, field experiments, trial deployment and implementation) and (ii) to balance carefully the technical, legal, ethical, economic and social concerns in a policy and governance framework, which is 'international in scope and remains flexible in light of fresh evidence' (The Royal Society 2009). If the first domain suggests that we are in the sphere of geoengineering science, here we feel we are crossing into the sphere of geoengineering politics:

> As geoengineering is considered more seriously, the question of norms to govern deployment will arise. Norms might be needed not only to determine when such systems might be used but also the kinds of evaluations that geoengineers might be required to make before deployment, compensation for parties harmed, cost sharing, and commitments to maintain geoengineering systems once deployed (Victor 2008, p. 330).

In fact, in this second domain the debate tends to move from the functioning of science to its interactions with policy and society, particularly by exploring three major narratives:

(i) The narrative of ethics, which introduces the questions of public value into the geoengineering field to overcome the difficulties of ensuring 'citizen representation' and the concerns of legitimacy associated with this (thus providing the basis for discussing the permissibility of the most controversial schemes and becoming an important legitimising factor for geoengineering activities). In the debate on geoengineering regulation and governance, this narrative focuses primarily on the concepts of fairness and justice, drawing upon formulations of environmental ethics and ethical and legal guidelines for human and animal subjects research (Keith 2002; Morrow et al. 2009; Miller 2010a):

    Yet, for me, phrases like "legitimate international process" and "all stakeholders" sound too much like climate scientists and government diplomats getting together to decide the fate of the planet. That hasn't worked so well so far, and not only because vulnerable developing countries have not been adequately consulted. So what kind of governance process do we need? To my mind, a potentially potent analogy is that of informed consent in human subjects research. Just like geoengineering research, human subjects research brings potentially significant public and private benefits by alleviating disease, injury, and even death. Yet, because such research is also very dangerous, societies have adopted strict regulations for the conditions under which that research can be done (Miller 2010a).

(ii) The narrative of failure, which spotlights the side effects and unintended consequences of geoengineering proposals, and therefore calls for the adoption of precautionary approaches and global, transparent and effective control and regulatory mechanisms. This narrative informs the politics of geoengineering, being particularly evident in the discussion of the risks surrounding field experiments with such technologies, concerns about unilateral attempts to conduct large-scale geoengineering actions, and the way they may weaken conventional mitigation and adaptation efforts, in what is referred to as the 'moral hazard' argument:

> In the context of geoengineering, the risk is that major efforts in geoengineering may lead to a reduction of effort in mitigation and/or adaptation because of a premature conviction that geoengineering has provided 'insurance' against climate change (The Royal Society 2009, p. 37).

(iii) Lastly, the 'valid science' narrative, suggesting possible ways to promote further 'strategic research' in the geoengineering field, the establishment of appropriate institutions for geoengineering governance, and greater citizen involvement, and calling for climate change science to become more critically reflective about its own role and impact. The 'valid science' narrative, appears under different forms on discourses of geoengineering regulation and governance, being particularly prominent in the debate on the involvement of relevant international scientific organisations, the establishment of international bodies and the first attempts to devise possible configurations to govern the research and deployment of geoengineering technologies (Olson 2011; Bodansky 2011, 2013; Bracmort and Lattanzio 2013).

> Meaningful research may also require actual trial deployment of geoengineering systems so that norms are informed by relevant experience and command respect through use. Standard methods for international assessment organized by the Intergovernmental Panel on Climate Change (IPCC) are unlikely to yield useful evaluations of geoengineering options because the most important areas for assessment lie in the improbable, harmful, and unexpected side effects of geoengineering, not the 'consensus science' that IPCC does well (Victor 2008, p. 321).

Together, these intertwined narratives tend to invoke the concept of 'good governance', which refers to the 'principles of openness, participation, accountability, effectiveness and coherence', and the need for science to function properly, i.e. 'assuring the productive functioning of its endeavours, and the maintenance of scientific integrity' (Siune et al. 2009).

### 11.3.3 Geoengineering Implementation and Misuse

The third domain includes the discussions surrounding the benefits and risks of using geoengineering to counteract global warming. While on the one hand we are still in the domain of empirical science, surrounded by simulation models that seek to address the climatic consequences of geoengineering schemes (Matthews and Caldeira 2007; Lenton and Vaughan 2009) and attempts to assess and rank different geoengineering methods – in terms of efficiency, affordability, safety, controllability, timeliness, reversibility, among others (Boyd 2008; Bellamy et al. 2012; The Royal Society 2009; Vaughan and Lenton 2011) – on the other hand we are also in the domain of 'geoengineering plausibility', where expectations, fears, fantasies, beliefs, and, of course, scientific expertise conspire to produce visions of *geoengineered worlds*.

The discourses about geoengineering implementation range from expert reviews that examine the potential advantages, drawbacks and risks of the different schemes to recent participatory processes that seek to elicit public and/or stakeholder views and perceptions of geoengineering (Bellamy et al. 2012). In these discourses we find many of the narratives previously identified, now being used to fill the empty spaces left by the inherent uncertainties associated with geoengineering technologies and climate change science. The positions vary and are not consistent with the traditional divisions usually found in the climate change debate—a 'quality' of the geoengineering debate that was already stressed by Jamieson in 1996:

> The recent debate makes for strange bedfellows. Many of those who believe most strongly that climate change is occurring are reluctant to embrace geoengineering approaches to reversing it. This is because they believe that the 'hand of man' is implicated in most of our environmental problems and they see geoengineering as more of the same. Others, who are interested in exploring or developing geoengineering possibilities, are disinclined to believe that climate is changing. On their view planetary systems are relatively insensitive to human behaviour and for that reason we shouldn't worry too much about the risks of geoengineering. So to simplify: some people believe that there is a problem but that geoengineering is no solution; others believe that geoengineering is a solution but that there is no problem (Jamieson 1996, p. 323).

This quotation is one of the many that emphasise the ambiguity and lack of correspondence between the various grey shades of geoengineering positions and the common black/white division between climate 'alarmists' and climate 'sceptics', which reinforces the importance of looking at geoengineering holistically.

Such a look is particularly relevant in examining the discourses about the potential misuses of geoengineering technologies, which should be considered in the broader context of the history of weather and climate modification, where many of the attempts to advance these technologies did not have peaceful intentions (Keith

2000; Fleming 2006, 2007, 2010; Bonnheim 2010).[7] In fact, the narratives about the risk of hostile uses of geoengineering technologies should be seen not only in the context of climate modification history, but also in the context of other potential harmful technologies:

> It may be possible to reduce the risk of intentional misuse through governance arrangements such as those that have been used to control nuclear, biological and chemical weapons. Similarly, it may be possible to prevent risks from unintentional misuse through sound regulation. However, in some cases the only effective measures for reducing risk may also forestall beneficial uses of geoengineering, for example by having a general chilling effect on scientific progress in this area (Powell et al. 2010, p. 2).

## 11.4 Concluding Remarks: Furthering the 'Democratisation and De-alienation' of the Geoengineering Debate

Using the internet as a primary source of information, we started our investigation by collecting different kinds of materials, including scientific articles, books, policy reports, films, interviews, media news and blog comments. We then analysed them to identify the main debates around geoengineering technologies.

Through this analysis we have identified three interconnected areas of current debate on geoengineering: geoengineering research and experimentation; geoengineering regulations and governance; and geoengineering implementation and misuse (Fig. 11.1). Within these areas we have also identified the main arguments called into question in the geoengineering debates and the underlying 'master narratives' in which they are embedded (Fig. 11.2).

The significance of geoengineering proposals can only be grasped in the context of the wider 'imaginary' of science and technology in which geoengineering narratives are rooted. Hence, we suggest examining those debates further, taking into consideration the dominant narratives of science, technology and society.

Different conceptions, understandings and value assumptions concerning the changing relationships between science and society, science and technology, and science and nature tend to shape the geoengineering debate and inform the analytical framework within which the geoengineering domain has been problematised (Scholte et al. 2013; Sikka 2012; Huttunen and Hildén 2013; Luokkanen et al. 2013; Nerlich and Jaspal 2012). This reinforces the need to unbind geoengineering discourses from the deeply embedded narratives of science, technology and society that present technoscientific innovation as the solution to our most critical problems and as a substitute for social change. Similarly, the construction of narratives that give meaning to human action within nature, and provide guidance for humans'

---

[7] Leading to the creation of new international legal instruments, such as the 1976 Convention on the Prohibition of Military or Any Other Hostile Use of Environmental Modification Techniques (ENMOD), and Articles 35(3) and 55(1) of the 1977 Additional Protocol I to the Geneva Convention.

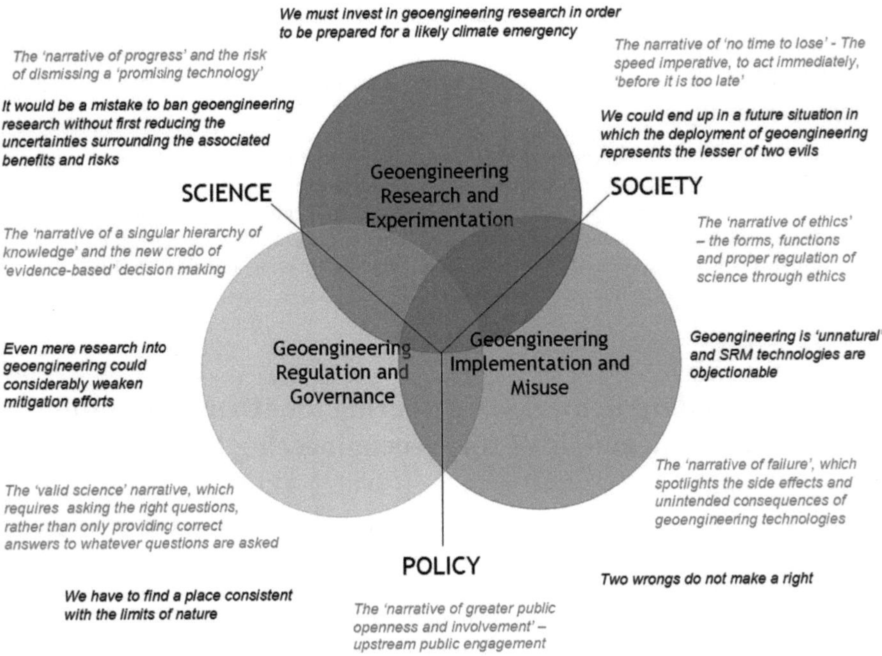


**Fig. 11.2** The three interconnected domains in which geoengineering debates are taking place: main arguments and underlying master narratives

domination of nature, deserves a more critical and open reflection than has been the case to date. As a result, many authors have been highlighting how important it is to consider public perceptions of geoengineering and therefore to help reveal the perceived moral orders underlying geoengineering proposals (Boyd 2008, Bracmort and Lattanzio 2013; Cicerone 2006, Miller 2010b; The Royal Society 2009).

The need for democratic decision-making and public engagement in the area of geoengineering has been clear for some time now (Jamieson 1996). However, only recently have the practical implications and challenges of such demands begun to be properly considered (Morton 2007; Miller 2010a, b; Powell et al. 2010; Bracmort and Lattanzio 2013; ETC Group 2010; NERC 2010a; Orr et al. 2011; Parkhill and Pidgeon 2011; Macnaghten and Owen 2011; Corner et al. 2012; Poumadère et al. 2011).

In the context of current 'policy vacuums', characterised by 'a growing sense of urgency coupled with a lack of knowledge of what to do and a lack of institutions where the issues could be addressed' (Rommetveit et al. 2010), these initiatives assume critical importance. And though the scope, scale and complexity of the climate change issues tend to 'render the fulfilment of the deliberative ideal a practical impossibility' (*Idem*), the recurrent claims that argue for a closer connection between science and society, with the purpose of exposing to public scrutiny the hidden assumptions, values and visions that are deeply embedded in geoengineering proposals, seem more than justified.

Therefore, in supporting the need to subject the scientific debate on geoengineering to more open and critical reflection, we highlight the importance of rebuilding the 'geoengineering scientific worldview' on social processes of trust and credibility (Irwin and Wynne 1996), in this way impelling climate change science to better reveal the competing interests, values and assumptions of climate engineering proposals. We also see this as an opportunity to promote critical thinking about social problems that tend to be 'circumvented' and reduced to technological fixes (Weinberg 1991), thus 'alienating' and 'diverting' our attention from an essential question, that of our place in nature.

In fact, the debates on climate engineering seem to offer an excellent framework within which to examine how modern science's 'alienation from the earth' is leading to the 'alienation from the world' (Arendt 1958), a condition clearly depicted by Funtowicz and Strand:

> Barring and bracketing the environmentalist talk – which also has been an important part of our own talk – of planetary dangers, we would like to propose that the planet is indeed not the object at risk. The object at risk is we ourselves as a collective (present and future) subjectivity and agency: the human right behind the human rights: that of personhood and hope. With personhood and hope in focus, the challenge is not the usual of what to do but, more importantly, how to do it as certain avenues of action are now deemed unacceptable (Funtowicz and Strand 2011, p. 8).

## References

Adam, David. 2009. Five extreme ways to beat climate change. *The Guardian*. http://www.guardian.co.uk/environment/2009/nov/04/extreme-climate-change-solutions/print.

Arendt, Hannah. 1958. *The human condition*, 2nd ed. Chicago: The University of Chicago Press.

Barrett, Scott. 2008. The incredible economics of geoengineering. *Environmental and Resource Economics* 39(1): 45–54.

Bellamy, Rob, Jason Chilvers, Naomi E. Vaughan, and Timothy M. Lenton. 2012. A review of climate geoengineering appraisals. *Wiley Interdisciplinary Reviews: Climate Change* 3(6): 597–615. doi:10.1002/wcc.197.

Bensaude-Vincent, Bernadette, Sacha Loeve, Alfred Nordmann, and Astrid Schwarz. 2011. Matters of interest: The objects of research in science and technoscience. *Journal for General Philosophy of Science* 42(2): 365–383. doi:10.1007/s10838-011-9172-y.

Betz, Gregor, and Sebastian Cacean. 2012. *Ethical aspects of climate engineering*. Karlsruhe: KIT Scientific Publishing.

Blackstock, Jason J., and Jane C.S. Long. 2010. The politics of geoengineering. *Science* 327(5965): 527. doi:10.1126/science.1183877.

Blackstock, Jason J., David S. Battisti, Ken Caldeira, Douglas M. Eardley, Jonathan I. Katz, David W. Keith, Aristides A. N. Patrinos, Daniel P. Schrag, Robert H. Socolow, and Steven E. Koonin. 2009. *Climate engineering responses to climate emergencies*. Novim Study Group, 2009 [cited 20.09.2010]. Available from http://arxiv.org/ftp/arxiv/papers/0907/0907.5140.pdf.

Bodansky, Daniel. 2011. Governing climate engineering: Scenarios for analysis. *Discussion paper 2011–47*, Harvard Project on Climate Agreements, Belfer Center for Science and International Affairs.

Bodansky, Daniel. 2013. *The who, what, and wherefore of geoengineering governance*. SSRN 2012 [cited May 11 2013]. Available from http://ssrn.com/abstract=2168850.

Bodle, Ralph. 2013. Climate law and geoengineering. In *Climate change and the law*, ed. Erkki Johannes Hollo, Kati Kulovesi, and Michael Mehling, 447–471. Dordrecht: Springer.

Bonnheim, Noah Byron. 2010. History of climate engineering. *Wiley Interdisciplinary Reviews: Climate Change* 1(6): 891–897.

Boucher, O., N. Gruber, and J. Blackstock. 2011. Summary of the synthesis session In *IPCC expert meeting report on geoengineering*, ed. O. Edenhofer, C. Field, R. Pichs-Madruga, Y. Sokona, T. Stocker, V. Barros, Q. Dahe, J. Minx, K. Mach, G.-K. Plattner, S. Schlömer, G. Hansen, and M. Mastrandrea. IPCC Working Group III Technical Support Unit, Potsdam Institute for Climate Impact Research, Potsdam, Germany.

Boyd, Philip W. 2008. Ranking geo-engineering schemes. *Nature Geoscience* Nov, 1: 722–724. doi:10.1038/ngeo348.

Bracmort, Kelsi, and Richard K. Lattanzio. 2013. Geoengineering: Governance and Technology Policy (Congressional Research Service Report No. R41471). Available from http://www.fas.org/sgp/crs/misc/R41371.pdf.

Bronson, Diana, Pat Mooney, and Kathy Jo Wetter. 2009. *Retooling the planet? Climate chaos in the geoengineering age*. A report prepared by ETC Group for the Swedish Society for Nature Conservation. Stockholm: Swedish Society for Nature Conservation. Available from http://www.etcgroup.org/content/retooling-planet-new-etc-group-report-geoengineering.

Brovkin, Victor, Vladimir Petoukhov, Martin Claussen, Eva Bauer, David Archer, and Carlo Jaeger. 2009. Geoengineering climate by stratospheric sulfur injections: Earth system vulnerability to technological failure. *Climatic Change* 92(3): 243–259.

Budyko, Mikhail Ivanovich. 1977. *Climatic changes*. Gidrometeoizdat, Leningrad (R) (1977) English translation by American Geophysical Union, Washington, DC, 261 pp.

Bunzl, Martin. 2009. Researching geoengineering: Should not or could not? *Environmental Research Letters* 4(4): 1–3. doi:10.1088/1748-9326/4/4/045104.

Carrier, Martin. 2011. Knowledge, politics, and commerce: Science under the pressure of practice. In *Science in the context of application*, ed. Martin Carrier and Alfred Nordmann, 11–30. Dordrecht: Springer.

Cicerone, Ralph J. 2006. Geoengineering: Encouraging research and overseeing implementation. *Climatic Change* 77: 221–226. doi:10.1007/s10584-006-9102-x.

Clark, William C., Hans J. Schellnhuber, and Paul J. Crutzen. 2004. Science for global sustainability: Toward a new paradigm. In *Earth system analysis for sustainability*, ed. Hans J. Schellnhuber, Paul J. Crutzen, William C. Clark, Martin Claussen, and Hermann Held, 1–28. Cambridge, MA: MIT Press.

Corner, A., N. Pidgeon, and K. Parkhill. 2012. Perceptions of geoengineering: Public attitudes, stakeholder perspectives, and the challenge of 'upstream' engagement. *Wiley Interdisciplinary Reviews: Climate Change* 3(5): 451–466. doi:10.1002/wcc.176.

Crutzen, Paul J. 2006. Albedo enhancement by stratospheric sulfur injections: A contribution to resolve a policy dilemma? *Climatic Change* 77(3–4): 211–220.

Crutzen, Paul J., and Eugene F. Stoermer. 2000. The anthropocene. *Global Change Newsletter* 41: 12–13.

Eccleston, Charles H., and Frederic March. 2011. *Global environmental policy: Concepts, principles and practice*. Boca Raton: CRC Press.

ETC Group. 2010. *Geopiracy – The case against geoengineering*. Ottawa: ETC Group. Available from http://www.cbd.int/doc/emerging-issues/etcgroup-geopiracy-2011-013-en.pdf.

Felt, U., B. Wynne, M. Callon, M.E. Gonçalves, S. Jasanoff, M. Jepsen, Pierre-Benoît Joly, Z. Konopasek, S. May, C. Neubauer, A. Rip, K. Siune, A. Stirling, and M. Tallacchini. 2007. *Taking European knowledge society seriously*. Report of the expert group on science and governance to the science, economy and society directorate, Directorate-General for Research, European Commission, Luxembourg.

Fleming, James R. 2006. The pathological history of weather and climate modification: Three cycles of promise and hype. *Historical Studies in the Physical and Biological Sciences* 37(1): 3–25.

Fleming, James R. 2007. The climate engineers: Playing god to save the planet. *Wilson Quarterly*, Spring 2007 31: 46–60.
Fleming, James R. 2010. *Fixing the sky – The checkered history of weather and climate control, Columbia studies in international and global history*. New York: Columbia University Press.
Fountain, Henry. 2003. The science of rain-making is still patchy. *New York Times* (October 19, 2003).
Funtowicz, Silvio O., and Jerome R. Ravetz. 1993. Science for the post-normal age. *Futures* 25(7): 739–755. doi:10.1016/0016-3287(93)90022-L.
Funtowicz, Silvio O., and Jerome R. Ravetz. 1994a. Emergent complex systems. *Futures* 26(6): 568–582. doi:10.1016/0016-3287(94)90029-9.
Funtowicz, Silvio O., and Jerome R. Ravetz. 1994b. Uncertainty, complexity and post-normal science. *Environmental Toxicology and Chemistry* 13(12): 1881–1885.
Funtowicz, Silvio O., and Roger Strand. 2011. Change and commitment: Beyond risk and responsibility. *Journal of Risk Research* 14(8): 995–1003. doi:10.1080/13669877.2011.571784.
Gardiner, Stephen. 2010. Is 'arming the future' with geoengineering really the lesser evil? Some doubts about the ethics of intentionally manipulating the climate system. In *Climate ethics, essential readings*, ed. S.M. Gardiner, S. Caney, D. Jamieson, and H. Shue, 284–312. Oxford: Oxford University Press.
Gardiner, Stephen. 2011. *A perfect moral storm: The ethical tragedy of climate change*. Oxford: Oxford University Press.
Gardiner, Stephen. 2012. The desperation argument for geoengineering. *PS: Political Science and Politics*. Available from http://ssrn.com/abstract=2183457.
Goodell, Jeff. 2010. *How to cool the planet: Geoengineering and the audacious quest to fix earth's climate*. Boston: Houghton Mifflin Harcourt.
Greene, C., B. Monger, M. Huntley. 2010. Geoengineering: The inescapable truth of getting to 350. *Solutions* 1(5): 57–66. Available from, http://www.thesolutionsjournal.com/node/771.
Hamilton, Clive. 2010. The powerful coalition that wants to engineer the world's climate. *The Guardian*. Available from, http://www.guardian.co.uk/environment/2010/sep/13/geoengineering-coalition-world-climate.
Hamilton, Clive. 2013. No, we should not just 'at least do the research'. *Nature* 496(7444): 139.
Haraway, Donna. 1997. *Modest_Witness@Second_Millennium.FemaleMan©_Meets_Oncomouse™, Feminist and Technoscience*. New York: Routledge.
Hegerl, Gabriele C., and Susan Solomon. 2009. Risks of climate engineering. *Science* 325: 955–956. Available from http://www.sciencemag.org/cgi/content/full/325/5943/955.
Huttunen, Suvi, and Mikael Hildén. 2013. Framing the controversial: Geoengineering in academic literature. *Science Communication*, June 28. doi:10.1177/1075547013492435.
Inman, Mason. 2010. Planning for plan B. *Nature* 4 (Nature reports – Climate change): 7–9. Available from http://www.nature.com/climate/2010/1001/full/climate.2010.135.html.
Irwin, Alan, and Wynne Brian. 1996. Introduction. In *Misunderstanding science? The public reconstruction of science and technology*, ed. Irwin Alan and Wynne Brian, 47–64. Cambridge: Cambridge University Press.
Jamieson, Dale. 1996. Ethics and intentional climate change. *Climatic Change* 33: 323–36.
Jasanoff, Sheila. 2004. *States of knowledge: The co-production of science and social order*. London: Routledge.
Jasanoff, Sheila. 2010. A new climate for society. *Theory, Culture & Society* 27(2–3): 233–253. doi:10.1177/0263276409361497.
Keith, David W. 2000. Geoengineering the climate: History and prospect. *Annual Review of Energy and the Environment* 25: 245–284.
Keith, David W. 2002. Geoengineering. In *Encyclopedia of global change*, ed. A.S. Goudie, 495–502. New York: Oxford University Press.
Keith, David W., and Hadi Dowlatabadi. 1992. A serious look at geoengineering. *Eos, Transactions American Geophysical Union* 73(27): 289–293.

Kintisch, Eli. 2010. The latest on geoengineering. *Science* 327(5969): 1070–1071. doi:10.1126/science.327.5969.1070-b.

Latour, Bruno. 1987. *Science in action: How to follow scientists and engineers through society*. Cambridge, MA: Harvard University Press.

Law, John. 2002. *Aircraft stories: Decentering the object in technoscience*. Durham: Duke University Press.

Lawrence, Mark G. 2006. The geoengineering dilemma: To speak or not to speak. *Climatic Change* 77: 245–248.

Lenton, T.M., and N.E. Vaughan. 2009. The radiative forcing potential of different climate geoengineering options. *Atmospheric Chemistry and Physics* 9: 5539–5561. doi:10.5194/acp-9-5539-2009.

Lovelock, James. 2008. A geophysiologist's thoughts on geoengineering. *Philosophical Transactions of the Royal Society A: Mathematical, Physical and Engineering Sciences* 366(1882): 3883–3890. doi:10.1098/rsta.2008.0135.

Lovelock, James. 2009. *The vanishing face of Gaia: A final warning*. London: Penguin.

Luokkanen, Matti, Suvi Huttunen, and Mikael Hildén. 2013. Geoengineering, news media and metaphors: Framing the controversial. *Public Understanding of Science*. doi:10.1177/0963662513475966.

MacCracken, Michaels C. 2009. *Beyond mitigation: Potential options for counter-balancing the climatic and environmental consequences of the rising concentrations of greenhouse gases*. Washigton, DC: The World Bank. Available from https://openknowledge.worldbank.org/handle/10986/9051.

Macnaghten, Phil, and Richard Owen. 2011. Environmental science: Good governance for geoengineering. *Nature* 479(7373): 293. doi:10.1038/479293a.

Marchetti, Cesare. 1977. On geoengineering and the CO2 problem. *Climatic Change* 1(1): 59–68. doi:10.1007/BF00162777.

Matthews, H. Damon, and Ken Caldeira. 2007. Transient climate–carbon simulations of planetary geoengineering. *Proceedings of the National Academy of Sciences (PNAS)* 104(24): 9949–9954. doi:10.1073/pnas.0700419104.

Michael, Mike. 2006. *Technoscience and everyday life: The complex simplicities of the mundane*. Maidenhead: Open University Press.

Miller, Clark A. 2010a. *The human subjects of geoengineering*. Arizona State University, Consortium for Science, Policy & Outcomes, [cited 08.09.2010]. Available from http://www.cspo.org/soapbox/view/100311P9IA/the-human-subjects-of-geoengineering/.

Miller, Clark A. 2010b. *International review of geoengineering*. Arizona State University, Consortium for Science, Policy & Outcomes, [cited 08.09.2010]. Available from http://www.cspo.org/soapbox/view/100319P4WK/international-review-of-geoengineering/.

Montenegro, Maywa, and Veronique Greenwood. 2009. *Will the future be geoengineered?* Available from http://seedmagazine.com/content/article/will_the_future_be_geo-engineered1/.

Morrow, David R., Robert E. Kopp, and Michael Oppenheimer. 2009. Toward ethical norms and institutions for climate engineering research. *Environmental Research Letters* 4. doi:10.1088/1748-9326/4/4/045106.

Morton, Oliver. 2007. Climate change: Is this what it takes to save the world? *Nature* 447: 132–136. doi:10.1038/447132a.

National Research Council (Panel on Policy Implications of Greenhouse Warming). 1992. *Policy implications of greenhouse warming: Mitigation, adaptation, and the science base*. Washington, DC: National Academy Press.

National Research Council. 2003. *Critical issues in weather modification research*. Committee on the status and future directions in U.S. weather modification research and operations. Washington, DC: National Academy Press.

NERC. 2010a. *Experiment earth? Report on a public dialogue on geoengineering*. The Natural Environment Research Council, The Sciencewise Expert Resource Centre, The Royal Society, and the multi-agency Living with Environmental Change (LWEC) programme.

NERC. 2010b. *Experiment earth – Public dialogue on geoengineering* (Geoengineering online survey from 5th March to 12th April 2010 – Summary report). Available from http://www.ipsos-mori.com/Assets/Docs/Publications/geoengineering_online-survey_final-report_june2010.pdf.

Nerlich, B., and R. Jaspal. 2012. Metaphors we die by? Geoengineering, metaphors, and the argument from catastrophe. *Metaphor and Symbol* 27(2): 131–147.

Olson, Robert L. 2011. *Geoengineering for decision makers*. Washington, DC: Woodrow Wilson International Center for Scholars. Available from http://www.wilsoncenter.org/event/report-release-geoengineering-for-decision-makers.

Orr, P.R., C.L. Twigger-Ross, E. Kashefi, K. Rathouse, and J.D. Haigh. 2011. *Experiment earth? Report on a public dialogue on geoengineering*. A report to the Natural Environment Research Council (NERC). Available from http://www.nerc.ac.uk/about/consult/geoengineering-evaluation-report.pdf.

Parkhill, Karen, and Nick Pidgeon. 2011. *Public engagement on geoengineering research: Preliminary report on the SPICE deliberative workshops* (Understanding risk group working paper, 11–01, Cardiff University).

Pielke, Roger Jr. 2010. *The climate fix: What scientists and politicians won't teel you about global warming*. New York: Basic Books.

Poumadère, M., R. Bertoldo, and J. Samadi. 2011. Public perceptions and governance of controversial technologies to tackle climate change: Nuclear power, carbon capture and storage, wind, and geoengineering. *Wiley Interdisciplinary Reviews: Climate Change* 2(5): 712–727. doi:10.1002/wcc.134.

Powell, Russell, Steve Clarke, Mark Sheehan, Tom Douglas, Bennett Foddy, and Julian Savulescu. 2010. *The ethics of geoengineering* [Working Draft]. Available from http://www.practicalethics.ox.ac.uk/__data/assets/pdf_file/0013/21325/Ethics_of_Geoengineering_Working_Draft.pdf.

Rayner, Steve, Catherine Redgwell, Julian Savulescu, Nick Pidgeon, and Tim Kruger. 2009. *Memorandum on draft principles for the conduct of geoengineering research*. House of Commons Science and Technology Committee enquiry into The Regulation of Geoengineering.

Rayner, Steve, Clare Heyward, Tim Kruger, Nick Pidgeon, Catherine Redgwell, and Julian Savulescu. 2013. The Oxford principles. *Climatic Change*. doi:10.1007/s10584-012-0675-2.

Rickels, W., G. Klepper, J. Dovern, G. Betz, N. Brachatzek, S. Cacean, K. Güssow, Heintzenberg J., S. Hiller, C. Hoose, T. Leisner, A. Oschlies, U. Platt, A. Proelß, O. Renn, S. Schäfer, and Zürn M. 2011. *Large-scale intentional interventions into the climate system? Assessing the climate engineering debate*. Scoping report conducted on behalf of the German Federal Ministry of Education and Research (BMBF). Kiel: Kiel Earth Institute.

Robock, Alan. 2008. 20 reasons why geoengineering may be a bad idea. *Bulletin of the Atomic Scientists* 64(2): 14–18.

Robock, Alan. 2011. Bubble, bubble, toil and trouble: An editorial comment. *Climatic Change*. doi:10.1007/s10584-010-0017-1.

Rommetveit, Kjetil, Silvio Funtowicz, and Roger Strand. 2010. Knowledge, democracy and action in response to climate change. In *Interdisciplinarity and climate change: Transforming knowledge and practice for our global future*, ed. R. Bhaskar, C. Frank, K.G. Hoyer, P. Naess, and J. Parker, 149–163. Abingdon: Routledge.

Sarewitz, Daniel. 2004. How science makes environmental controversies worse. *Environmental Science and Policy* 7(2004): 385–403.

Schelling, Thomas. 1996. The economic diplomacy of geoengineering. *Climatic Change* 33(3): 303–307. doi:10.1007/BF00142578.

Schneider, Stephen H. 1996. Geoengineering: Could— or should— we do it? *Climatic Change* 33(3): 291–302.

Scholte, Samantha, Eleftheria Vasileiadou, and Arthur C. Petersen. 2013. Opening up the societal debate on climate engineering: How newspaper frames are changing. *Journal of Integrative Environmental Sciences* 1–16. doi:10.1080/1943815x.2012.759593.

Schwartz, Peter, and Doug Randall. 2003. *An abrupt climate change scenario and its implications for United States National Security* (report prepared by the Global Business Network (GBN) for the Department of Defence).

Schwarz, Astrid, and Wolfgang Krohn. 2011. Experimenting with the concept of experiment: Probing the epochal break. In *Science transformed? Debating claims of an epochal break*, ed. Alfred Nordmann, Hans Radder, and Gregor Schiemann, 119–134. Pittsburgh: University of Pittsburgh Press.

Sikka, T. 2012. A critical discourse analysis of geoengineering advocacy. *Critical Discourse Studies* 9(2): 163–175.

Singer-Vine, Jeremy. *Choose your own Geohack. Explore geoengineering's most interesting strategies with Slate's interactive guide.* The Slate Group [September 24, 2010]. Available from http://www.slate.com/id/2268479/.

Siune, K., E. Markus, M. Calloni, U. Felt, A. Gorski, A. Grunwald, A.Rip, V. Semir, and S. Wyatt. 2009. *Challenging futures of science in society – Emerging trend and cutting-edge issues. Report of the MASIS expert group setup by the European Commission.* Brussels: European Commission, Directorate-General for Research Science in Society.

The Royal Society. 2009. *Geoengineering the climate: Science, governance and uncertainty.* London: The Royal Society.

UK House of Commons. Science and Technology Committee. 2010. *The regulation of geoengineering, fifth report of session 2009–10.* Available from http://www.publications.parliament.uk/pa/cm200910/cmselect/cmsctech/221/221.pdf

Vaughan, Naomi, and Timothy Lenton. 2011. A review of climate geoengineering proposals. *Climatic Change* 109(3–4): 745–790. doi:10.1007/s10584-011-0027-7.

Victor, David G. 2008. On the regulation of geoengineering. *Oxford Review of Economic Policy* 24(2): 322–336.

Victor, David G., M. Granger Morgan, Jay Apt, John Steinbruner, and Katharine Ricke. 2009. The geoengineering option–A last resort against global warming? *Foreign Affairs* 88(2): 64–76.

Ward, Robert DeCourcy. 1930. How far can man control his climate? *The Scientific Monthly* 30: 5–18.

Weinberg, Alvin M. 1991. Can technology replace social engineering? In *Controlling technology: Contemporary issues*, ed. W.B. Thompson, 41–48. Buffalo: Prometheus Books.

Williamson, P., Watson, R.T., Mace, G., Artaxo, P., Bodle, R., Galaz, V., Parker, A., Santillo, D., Vivian, C., Cooper, D., Webbe, J., Cung, A., and E. Woods. 2012. Impacts of climate-related geoengineering on biological diversity. Part I of: *Geoengineering in relation to the convention on biological diversity: Technical and regulatory matters*. Secretariat of the Convention on Biological Diversity. Montreal, Technical Series No. 66, 152 pages.

WMO. 2010. *Executive summary of the WMO statement on weather modification* (Expert team on weather modification research, Abu Dhabi, 22–24 March 2010). Available from https://www.wmo.int/pages/prog/arep/wwrp/new/documents/WMR_documents.final_27_April_1.FINAL.pdf.

Yanarella, Ernest J., and Christopher Rice. 2011. Global warming and the specter of geoengineering: Ecological apocalypse, modernist hubris, and scientific-technological salvation in Kim Stanley Robinson's global warming trilogy. In *Engineering earth: The impacts of megaengineering projects*, ed. Stanley D. Brunn, 2233–2252. Dordrecht: Springer.

# Coda: Governing Technological Emergence in the Digital Society: Political Tensions and Ethical Dilemmas

The introduction of this book explores some concepts on the subject of governance that is the basis of various chapters and also serves as a link between them. This deliberate choice was not made to provide a ready-made or closed definition of governance. The empirical findings of this book do indeed suggest that the governing of science and technology is not predictable but that it takes different shapes and that it is enacted in different sets of practices within socio-technical and historical contexts. There is no singular definition of governance. Ethics should be recognized as an element of governance. Whereas governance is not a given, one can always think of *better* forms of governance. However, in scholarly literature, politics and ethics are often kept as separate domains. The contents of this book show that technology, politics and ethics are inevitably intertwined. Values are displayed in the practice of governance, and ethical dilemmas arise that need to be dealt with in democratic ways. The chapters of this book have combined approaches from STS and from ethics in an attempt to present governance as a set of practices, institutional sites, discourses, regulatory mechanisms and public responses that are not given, but are negotiated. Ethics is a regulating element in those negotiations as well as an effect of them. Only in this way could the governance of science be a democratic process.

While avoiding a single definition of governance, the introduction of the book provides some ideas on how governance is often understood as a working concept in scholarly and policy contexts. This conclusion revisits that discussion as a point of departure and as a way to make sense of the empirical findings of the book as a whole. As explained in the introduction, the book is a collective effort of a group of scholars working in a European context and collaborating on a project supported by the European Commission. In a European policy context, the emergence of technologies such as geo-engineering and smart ICT systems is often portrayed as the preferred formula to address grand challenges. It also emphasizes that these technologies are not only promising but also entail important potential risks. With this focus on potential, promises and risks, the governance of emerging technologies is



A. Delgado (ed.), *Technoscience and Citizenship: Ethics and Governance in the Digital Society*, The International Library of Ethics, Law and Technology 17,
DOI 10.1007/978-3-319-32414-2

then understood as being irremediably oriented towards the future and inhabited by the Collingridge dilemma:

> The social consequences of a technology cannot be predicted early in the life of the technology. By the time undesirable consequences are discovered, however, the technology is often so much a part of the whole economic and social fabric that its control is extremely difficult. This is the dilemma of control. When change is easy, the need for it cannot be foreseen; when the need for change is apparent, change has become expensive, difficult and time consuming. (Collingridge 1980:11 in Nordmann 2010).

The ethical concern behind the Collingridge dilemma is with the impossibility of having control over technological development, use, application and impacts. The diagnosed problem is, eventually, the lack of knowledge. In response, throughout the years, a number of technologies of governance have been developed in a European policy context, such as different forms of public participation, opinion polls, scenarios, foresight exercises, etc. As part of this policy context, the authors of this book received economic support from the European Commission (FP7 program) to collaborate in a project called Technolife. This book is a result of that project. Using movies to trigger debate, Technolife developed three online forums to facilitate discussions on emerging technologies that increasingly affect people's everyday lives. These technologies are digital maps, biometrics and human enhancement technologies (see the introduction to this volume). People whose everyday life was likely to be affected by the developments of such technologies[1] were invited to participate. Perhaps the most remarkable result from the forums was that although people also showed a concern with 'controlling technology', the concern was much less on risks and much more on politics. Many participants expressed concerns that were not so much on the type of uncertainties and risks that come with technological emergence (as in the Collingridge dilemma), but on who controls technology and the resulting effect on people's life. Many participants were critical of the monopolies of mass media and large corporations, and many participants saw politics and bureaucracies as out-dated institutions. While technology was often presented as having great potential for social change, institutions were seen as not being able to cope with changes:

> In this era of rapid and sweeping advancement, we see the old world struggling to guide and restrain the process of advancement into the new (next?) world. Recording companies howl bloody murder in the old courts about people 'stealing their livelihood' by making and distributing pirate copies of their intellectual properties. Yesterday's telephone companies become today's facilitators of information and entertainment access. World governments gnash their teeth at the possibility of new technologies sparking sweeping economic change and the dashing of the old world's entrenched economic power structures. Change will happen according to the will and abilities of the masses, regardless of the old world sensibilities (Body enhancement forum participant A)
>
> The only thing I am concerned about is if all of this would be affordable to common people. I don't care if someone doesn't want to improve memory or add years to life or

[1] Their occupation was a main aspect taken into consideration. For instance, for the body enhancement forum, invited participants were models, athletes, body builders…but also…

> technologically advance their body. I care if someone wants to do that but lacks money (BH, forum participant B).
>
> Right now, the biggest problem I see is the fact that these new technologies are being developed in a hyper-capitalistic environment and are being registered to pharmaceutical companies (referring to human enhancement technologies) (BH, forum participant C)
>
> I've never seen an institution with the slightest interest in improving our lives (BH, forum participant D).

Expressed distrust in institutions was often accompanied with (sometimes extreme) technological optimism, especially concerning the type of distributed social action enabled by ICTs:

> Push a little further down that line and we might see fully-automated virtual tools that let the layperson design unique organisms via their home computer and distribute the fruits of their labors to all interested parties across the globe with one tiny command. When that day comes, Big Pharma will compete against the ubiquity of information and the will of the people, and it will lose[2]

Being imagined as ubiquitous and decentralized, digital technologies are said to potentially enable people with economic and political autonomy. Somewhat ironically, it is the assumingly deregulated character of ICTs that may enable people to have more control (in the sense of independency), as the quote above indicates. The use of ICTs in different technological and non-technical domains was emphasized in these debates as posing political challenges that institutions were not prepared to address. They were frequently discussed as issues of privacy (vs security), openness, distributed ownership, and freedom (for enhancing bodies and minds, among other things). As they were articulated in the debates, concerns were repeatedly expressed as to who controls technology and by which means. In these discussions, the participants of these forums confronted an institutionalized way of thinking and practicing the governance of emerging technologies that is still too focused on uncertainly, risk and evidence to emphasize governance as a political matter. Furthermore, issues such as privacy and access to technology and information were articulated as 'claims to rights' (Ruppert and Isin 2015). The formulation of those issues as claims to rights can be seen as enacting confrontation against mass security systems, surveillance or restricted access, as well as a way of confronting governance as usually exercised. In other words, the participants of these forums were performing not as only *stakeholders* but also as *citizens* (through formulating *claims to rights*), as they are usually considered in policy contexts, particularly in institutionalized public engagement exercises where issues are often formulated as public *opinions, views or perceptions* (which can also be viewed as a method of distilling political agency).

The chapters of this book echo the concerns that emerged in the Technolife discussion forums. They provide empirical findings and discussions around the issue of "controlling technology" paying attention to ethics, politics and citizenship, as they appear in three contexts of technological emergence. As technological interventions on the body of citizens, their movements and the space they inhabit are

[2] All quotes from forum participants are extracted from the Technolife project final report.

explored throughout the book, a political tension becomes apparent. ICT based technologies, such as digital maps, might produce a perplexing effect. These technologies are often inserted in larger infrastructures and allow new forms of institutional control, as we see in the case of biometrics and the attempts at controlling citizen's movements across borders. For instance, the insistence on "interoperability" of biometrics systems would eventually enable an expanded control beyond national borders in the pursuit of a common European border. Therefore, interoperability turns out to be a technology of government. On one hand, digital platforms and devices may be seen as enabling a closer relation with technology, one in which people can to some extent exert a certain control over (and through) the technology, in their own lives and choices. This is not a straightforward effect because technologies can also produce an ambiguous feeling of being simultaneously gaining and losing control over one's own life. Chapter 3, on Cochlear implants, shows how people using this technology are described as needing to sacrifice some important dimensions of the way in which they experienced their life in order to learn how to live with an enabling technology.

Recurrent attempts at gaining control over technologies that affect people's live, reveal incipient ways of enacting values such as autonomy, privacy and freedom. As mentioned, these were values repeatedly invoked, discussed and claimed in the Technolife forums (interestingly, in all of the forums and in different technological fields). Sharing and open access were often invoked. Working within digital infrastructures, information appears to be more accessible and to circulate more openly, but digital applications can also be more easily tracked. The possibility of tracking the movements of citizens, combined with increasing systems interoperability, provides states with enhanced visibility and control capabilities. States cannot be autonomous in the development and implementation of such large scale technological systems because they are largely dependent on corporate interest. This tension between large, national and local scales, as well as between monopolies and distributed economic and political agencies, was mentioned many times in the forum and was explored in the section on biometrics. The findings from the forum suggest that the huge efforts deployed by governments to produce a technology of trust (to make Europeans feel safe in a "security envelope") may reveal a lack of trust. Thus, we could conclude with Marilyn Strathern that "a benevolent or moral visibility is all too easily shown to have a tyrannous side—there is nothing innocent about making the invisible visible (2000: 309)". What was eventually questioned in the section on biometrics was the type of visibility (and effects) that such technological systems and infrastructure deploy as they work to make the movement of citizens an object of transparency.

Finally, a tension that was recurrently referred to in the Technolife forums was the tension between the type of individual and collective political agencies entangled and produced within the processes of technological emergence. "Many share the commitment to values such as pluralism and the individuality of choice. What's more, many explicitly articulate these values in direct connection with a concern over standardized, top-down institutions, modes of production, distribution of goods, resources and information" (Strand and Rommetveit 2011). Technological decentralization, individual freedom and diversification were opposed to controlling governmental agencies that are known to impose regulation and standardization. However, the multiple criticism to monopolies and capitalism was difficult to interpret, as it could lean toward libertarianism, extreme liberalism (or indeed towards something not known yet). In the governance of emerging technologies, such as those explored in this book, that tension is performed within practices of information sharing. As in the sharing economy, broad circulation of information in networked platforms is expected to produce value. By using digital maps, people give away some information to then be able to make other choices. One example of this would be customizing spaces through apps in cell phones. The figure of the citizen as a consumer co-exists with collective forms of action enabled by technology, such as when people produce maps for monitoring their environment, or visualizations of polluted areas.

This conclusion summarizes some of the tensions that arise in the governance of emerging technologies that were main findings of the Technolife project and that are reiterated by the findings of this book. As these findings suggests, a main challenge is for institutions to respond to a real state of affairs and to address emerging technologies not only as a tool to cope with societal challenges but also as embedded within the constellations of socio-political relations, posing new societal challenges as they emerge. Today, ICTs are loaded with optimism and hope on the side of both institutions and the citizenry; however, in the pursuit of better forms of governance, the reconfiguration(s) of power relations that comes with technological emergence should not be concealed in extreme technological reliance.

Ana Delgado

## References

Nordmann, Alfred. 2010. A forensics of wishing: Technology assessment in the age of technoscience. *Poiesis and Praxis* 7(1): 5–15.

Ruppert, E., and E. Isin. 2015. *Being digital citizens*. London: Rowman and Littlefield International. ISBN 978-1-178348-055-5.

Strand, R., and K. Rommetveit. 2011. *Technolife project final report*. Retrieved from http://technolife.no/content/filelist_b9b0f429-0e6c-49f5-8944-24541635e46e/1336510179106/technolife_final_report_for_website.pdf.

Strathern, M. 2000. The tyranny of transparency. *British Educational Research Journal* 26(3): 309–321.